SpringerBriefs in Ethics

Springer Briefs in Ethics envisions a series of short publications in areas such as business ethics, bioethics, science and engineering ethics, food and agricultural ethics, environmental ethics, human rights and the like. The intention is to present concise summaries of cutting-edge research and practical applications across a wide spectrum.

Springer Briefs in Ethics are seen as complementing monographs and journal articles with compact volumes of 50 to 125 pages, covering a wide range of content from professional to academic. Typical topics might include:

- Timely reports on state-of-the art analytical techniques
- A bridge between new research results, as published in journal articles, and a contextual literature review
- A snapshot of a hot or emerging topic
- In-depth case studies or clinical examples
- Presentations of core concepts that students must understand in order to make independent contributions

Yoann Della Croce

Assisted Dying, Discrimination, and Structural Injustice

Springer

Yoann Della Croce
Faculty of Law
McGill University
Montreal, QC, Canada

ISSN 2211-8101 ISSN 2211-811X (electronic)
SpringerBriefs in Ethics
ISBN 978-3-032-11092-3 ISBN 978-3-032-11093-0 (eBook)
https://doi.org/10.1007/978-3-032-11093-0

This Springer imprint is published by the registered company Springer Nature Switzerland AG
The registered company address is: Gewerbestrasse 11, 6330 Cham, Switzerland

Acknowledgements

Albeit short, this book grew out of innumerable stimulating discussions going back to when I was a doctoral student writing a dissertation on the ethics of inequality of access to assisted suicide. As much of this book's content has been inspired by my doctoral work, much of the credit behind the refinement of its ideas must go to my dissertation supervisors, Matteo Gianni and Samia Hurst, as well as to my dissertation examination committee, Michael Cholbi, Soren Holm, and Frédéric Varone, whose comments have contributed immensely not only to my dissertation but also to most of the ideas present in this short manuscript. I am indebted to, in no particular order, Kasper Lippert-Rasmussen, Soren Midtgaard, Jonas-Sébastien Beaudry, Naïma Hamrouni, Ryoa Chung, Claudia Gamondi, Ophelia Nicole-Berva, and Esma Baycan-Herzog, as well as to audiences at the University of Geneva, McGill University, and the Centre de Recherche en Ethique for their many insightful comments which have definitely and drastically improved the quality of this manuscript. I am also thankful for the ideal writing environment I have been lucky to be granted access to at the University of Geneva, Aarhus University and McGill University, all of which have left their imprint on the ideas discussed here. I would also like to thank the Swiss National Science Foundation (SNSF), which has granted me the financial support needed to see this project to its end. Finally, as is the case with all pieces of academic research, the number of people who have in one way or another contributed to this book is too great to ever properly do justice to. Special thanks are nevertheless to be given to my family, friends, colleagues, and roommate: this manuscript is yours as much as it is mine.

Competing Interests The author has no competing interests to declare that are relevant to the content of this manuscript.

About the Book

While frequent, the use of the idea of discrimination in the literature on the ethics of assisted dying is usually conceptually unclear and at times almost seemingly rhetorical. This book provides an overview of the many usages of discrimination in the ethics of assisted dying and attempts to conceptually map these according to an analytical account of discrimination, differentiating between direct and indirect discrimination at the individual and organizational levels. Particular attention is dedicated to the case of indirect discrimination due to structural factors, which I argue has been significantly and systematically overlooked in the literature. While the literature has traditionally engaged with the discrimination on the basis of disability (i.e.; discrimination as disparate impact of legalizing assisted dying or discrimination as inequality of opportunity in accessing assisted dying) or of exclusionary legal barriers (i.e., forbidding the mentally ill from accessing assisted dying), comparatively little attention has been devoted to the cases of individuals who are not legally excluded from access, are physically able to commit suicide, but are yet unable to effectively access assisted suicide due to their situations of structural vulnerability. Such cases, for instance, include inhabitants of rural areas who have no reasonable access to relevant facilities or cases involving low-income individuals, amongst others. A conceptual mapping of discrimination in the ethics of assisted dying not only clarifies the kind of moral wrongs brought to the fore in current research on the subject but also provides a finer understanding of the overall architecture of the ethical debate. Moreover, taking discrimination seriously helps in casting light on the inherent political dimension of assisted dying through the relation between discrimination and structural injustice as well as vulnerability, particularly when looking at cases of indirect discrimination. Once this political dimension of assisted dying is demonstrated, it appears clear that the ethics of assisted dying are not only to be concerned with the principles of autonomy and liberty but also with those of justice and equality, adding to the claim that death and dying are issues of social justice. This in turn allows for the elaboration of a robust normative claim concerning the need to ensure fair and equal opportunity of access to assisted dying.

This approach reinforces the claim that the morality of assisted dying is not to be assessed through a binary answer regarding its permissibility or impermissibility, but rather that assisted dying laws exist on a continuum on which certain arrangements are most just than others and thus calls for further research in this direction.

Contents

1 Introduction . . . 1
References . . . 5

2 Discrimination and Assisted Dying . . . 7
2.1 Preliminary Conceptual Remarks on Discrimination, Assisted Suicide, and Euthanasia . . . 7
2.2 The Many Faces of Discrimination in Assisted Dying . . . 14
2.2.1 Disability . . . 14
2.2.2 Mental Disorders . . . 17
2.2.3 Existential Suffering . . . 21
2.2.4 Rural Communities . . . 23
2.2.5 Low-Income Groups . . . 29
2.3 Mapping Discrimination in Assisted Dying . . . 32
References . . . 36

3 Structural Injustice in Assisted Dying . . . 41
3.1 From Discrimination to Structural Injustice . . . 41
3.2 Three Occurrences of Structural Injustice in Assisted Dying . . . 44
3.2.1 Devaluation of Life and Expressivism . . . 46
3.2.2 Access to Palliative Care and Choosing When Choices Are Limited . . . 48
3.2.3 A Forgotten Face: Structural Barriers to Access . . . 50
References . . . 53

4 Conclusion: Politicizing Assisted Dying . . . 57
References . . . 61

Chapter 1
Introduction

Abstract This introductory chapter lays out the context in which this book was written and highlights the fictitious division between opponents of assisted dying and how this side of the debate properly accounts for background conditions of injustice as opposed to those in favor who allegedly champion individual liberty without much regard for such background. I lay out some of the key points of the book, one of which being to come to an end with this division by arguing that background conditions are relevant to all normative theorists working on the ethics of assisted dying and are not just to be used as argument by those who oppose the practice. I then present a brief overview of the current state of assisted dying legislations and its contemporary challenges. I conclude by presenting the structure of the book and introduce its different parts.

Keywords Assisted dying · Euthanasia · Assisted suicide · Liberalism · Injustice · Discrimination · Disability

When I was in the midst of the drafting of this short book, I recall one very specific instance that I believe illustrates well the motive behind its content. I was co-organizing a colloquium on new issues in the ethics of assisted dying and at some point, someone in the audience wanted to raise some criticism to one of our participant's contributions. I remember the words they used very well to start off their critique: "I am anti-ableism, anti-racism, anti-sexism, anti-colonialism, and therefore, anti-assisted dying". I was left confused by such words, for it had never occurred to me that there would be some sort of connection between endorsing ideals of social justice and opposing assisted dying. I thought to myself that I was after all not one of those hardcore libertarians that champion individual liberty over everything else, and that I was deeply concerned with all the injustices raised by this person. We were essentially fighting for the same thing. And yet, here I was defending expanding access to assisted dying, unknowingly giving off the impression that I actually was the hardcore libertarian I knew I had never been. Far from an isolated claim made by one person during a particular colloquium, I believe there is a widespread assumption that being concerned with social justice should at least make one

Y. Della Croce, *Assisted Dying, Discrimination, and Structural Injustice*, SpringerBriefs in Ethics, https://doi.org/10.1007/978-3-032-11093-0_1

highly suspicious of legalizing, let alone expanding, assisted dying. This is in fact not terribly surprising, for as we shall see shortly the bulk of arguments against assisted dying, casting aside those who find their origins in theological approaches, usually base their claims on pre-existing forms of structural injustice and oppression and how such dynamics may end up spawning catastrophic consequences if assisted dying was to be legalized (or expanded). This has given rise to a caricatural divide between those concerned with social justice who seek to limit the harm caused to vulnerable social groups by assisted dying on the one hand, and "color-blind" liberals who brandish the ideal of individual liberty carelessly on the other. If there is only one contribution that this short book aims to make, it is show that this divide is to be torn down. Those in favor of assisted dying can, and should, very well be concerned with structural injustice, oppression, and social justice particularly insofar as inequality of access is concerned. The pictures that ought to be painted by the ethics of assisted dying must feature more nuance and complexity for the debate risks, and admittedly at times has, becoming little more than a rhetorical joust between people with substantially different convictions but little actual philosophical substance to defend them. As such, this book's claims are conceptual, normative, at times methodological, as well as policy-oriented, for I aim to show that ideas such as that of structural injustice and discrimination ought to be taken seriously both by scholars of the field and policymakers. In order to do that, we must begin with a set of questions that will guide our present inquiry about structural injustice and discrimination in assisted dying. These questions are many. What do we mean when we say that *x* instance of unequal access to assisted dying is discriminatory? Or when we say that assisted dying is itself discriminatory to those in situations of social and political vulnerability? And what does such diagnoses of discrimination involve for those in charge of designing and putting laws into practice?

Discrimination is an oft-invoked concept in the ethics of assisted suicide and euthanasia. In the earlier days of the philosophical debate concerning the moral permissibility of legalizing some form of assisted dying, discrimination has been mostly put forward as an argument from the anti-legalization camp, claiming that legalization will disparately impact the most vulnerable members of society, particularly people with disabilities. Some of the most stringent critics have gone as far as claiming the legalization of assisted suicide to be the ultimate, and most lethal, form of discrimination against people with disabilities (Bickenbach 1998). The kind of discrimination at stake here echoes traditional slippery-slope arguments against the legalization of assisted dying on the basis of what falls under the heading of disparate impact also referred to as indirect discrimination, as opposed to disparate treatment, also referred to as direct discrimination (Moreau 2010). While this anti-legalization stance has not completely lost traction and is still supported by some (Coleman 2010; Gill 2010; Braswell and Garland-Thomson 2023), it has also witnessed its fair share of criticism for allegedly being too paternalistic and victimizing to the social group of people with disabilities (Silvers 1998; Riddle 2017; Della Croce 2020; Colburn 2022).

Recent years have however seen a slow but steady shift in how discrimination is utilized in the philosophical debate. A growing number of scholars now make use of the concept of discrimination not to substantiate anti-legalization claim but rather to question the moral legitimacy and justification of the boundaries of legislation; this is at least partially due to the growing number of jurisdictions that have adopted law that permits aid in dying in at least some form.[1] Assisted dying laws are now also understood as existing on a continuum and the ethical debate is now not only concerned with the permissibility of legalization but with which arrangements are most just and morally legitimate (Sleeman and Chalmers 2019); call this the *continuum approach* as opposed to the *dichotomic approach*. Perhaps the most prominent illustration of this new approach to discrimination is to be found in arguments that question the moral justifiability of excluding the mentally ill from accessing assisted dying (Schuklenk and van der Vathrost 2015; Varelius 2015; Rooney et al. 2018; Hatherley 2019). As opposed to the disparate impact or indirect discrimination sense seen above, this line of argument uses discrimination in its disparate treatment or direct discrimination sense: laws and policies that frame the practice of assisted dying are deemed downright exclusionary *by design* and not by unforeseen consequence.

Discrimination has also been used by various advocates for disability rights both on the continuum as well as on the dichotomic approach. Pro-legalization scholars have argued that some persons with disabilities effectively lack the physical capacity to end their own lives, thus placing them on unequal standing with regards to able-bodied citizens as they had no possibility to exercise their private liberty-right to suicide without assistance, hence the need for legalization (Sneddon 2006).[2] Others, on the continuum approach, have argued on similar grounds that current aid in dying laws were discriminatory because of the self-administration requirement, which *de facto* excluded otherwise-eligible patients suffering from degenerative neurologic diseases or neuromuscular disabilities (Shavelson et al. 2023). In line with this argument, it has also been argued that the self-administration requirement was not the only way in which assisted dying laws discriminated against people with disabilities, as the manner in which the communication and decisional capacity requirements are framed can also unfairly exclude persons with cognitive disabilities as well as communication disorders from accessing the procedure (Wright 2023).

It is worth noting that the normative value ascribed to assisted dying varies depending on the kind of discrimination-based argument that is brought up and who it is brought up by. Looking at the anti-legalization camp, discrimination is used in order to highlight the unfair distribution of burdens that the legalization of assisted

[1] At the time of writing, assisted dying (be it through assisted suicide or euthanasia) is legal under some circumstances in Belgium, Canada, Colombia, Luxemburg, the Netherlands, New Zealand, Spain, Switzerland, the US states of California, Colorado, Hawaii, Maine, Montana, New Jersey, New Mexico, Oregon, Vermont, Washington and Washington D.C. as well as in the six states of Australia (New South Wales, Queensland, South Australia, Tasmania, Victoria and Western Australia) (Montagna et al. 2023).

[2] This line of reasoning is at the heart of both the *Rodriguez v. British Columbia* as well as the *Carter v. Canada* cases, the latter being the turning point that allowed for the legalization of medical aid in dying in Canada.

suicide may bring along; assisted suicide is thus conceptualized as some sort of evil that will impact more heavily on the vulnerable. These arguments presume some sort of badness inherent to assisted dying. Arguments on the continuum approach in favor of non-discriminatory laws, on the other hand, conceptualize assisted suicide as a right that should be able to be realized by legitimate requestors. Contrary to what is put forward by the anti-legalization camp, these arguments presume some sort of goodness of assisted suicide. It is however true that arguments pertaining to the continuum approach do not necessarily assume that assisted suicide is good, nor are they necessarily in favor of expanding access to assisted suicide. One can easily imagine an argument that, while not questioning the legitimacy of assisted suicide for terminally ill individuals, rejects the idea of access to those with mental disorders. As we will see later, this does not prove relevant for the analysis at stake here. This is because the use of discrimination-based arguments is made solely by those in favor of expanding access on the continuum approach, which is intuitive. The normatively ascribed categories I bring up here are only valid insofar as discrimination-based arguments in assisted suicide are concerned and should not be transposed to the dichotomic vs continuum distinction: they are only to be understood as analytical constructs that make sense within the realm of the present inquiry about discrimination.

There are many other ways in which discrimination is brought up in the literature on assisted dying, the ones presented above being just illustrations of perhaps the most salient and straightforward usages of the concept in the debate; we will turn to other cases later in the analysis. There however appears to be a discrepancy with regards to how often the idea of discrimination is brought up compared to how often assisted dying scholars have attempted to clearly define what exactly the mean by discrimination and more specifically what kind of discrimination is at stake in their respective arguments. As it has been highlighted above, discrimination can be used as a normative weapon to argue for both anti and pro legalization on the dichotomic approach as well as a tool to criticize the boundaries of existing legislations on the continuum approach; it effectively is used indiscriminately by all voices in the debate. Using this observation as a starting point, this book divides itself into two main sections: a first conceptual section that aims at clarifying the usages of discrimination in the ethics of assisted dying, and a latter normative section that argues about the benefits of this conceptual mapping as far as social justice is concerned. First, I show that discrimination can be identified in other cases than the more evident aforementioned ones, particularly insofar as it relates to structural factors that impair access to assisted dying. This, along with a systematic review of the other uses of discrimination and cases of inequality of access within the debate, allows for an exhaustive listing of instances of discrimination to assisted dying procedures. Second, I propose a conceptual mapping of instances of discrimination in assisted dying based on Lippert-Rasmussen's analytical account of discrimination (Lippert-Rasmussen 2014). The identified instances are mapped onto the categories of direct discrimination, indirect discrimination, and nondiscrimination. Cases are then further categorized according to the *locus* of discrimination, legal or structural, and the moral justifiability of the exclusion. This mapping allows for a clear conceptual overview of the ways in which discrimination is used and provides guidance in identifying the sources of discrimination. It is important to note that my aim here is not to provide a critical assessment of each of the uses of discrimination in assisted dying as the goal of this book is not to determine the validity of each position. Rather, my aim is to simply present each position as charitably as possible and determine if the use of discrimination is indeed warranted, and if it is what kind of discrimination each position alludes to. Once these conceptual clarifications have been made, I tackle the subject of how structural injustice interfaces with discrimination in access to assisted dying. After defining structural injustice and laying out its conceptual relation with discrimination and vulnerability, I review some typical argumentative instances often present in the ethics of assisted dying and show how they, although often implicitly, rely on the idea of structural injustice to make their claims. While these arguments usually rely on structural injustice to argue against assisted dying, I show that structural injustice can also be used to highlight deficiencies in equality of access. I then conclude by emphasizing the importance of thinking in terms of structural injustice to ensure a proper politicization of assisted dying and highlight why such politicization is crucial to the future of the literature, be it in terms of how research on the subject, both theoretical and empirical, should be conducted as well as in what kinds of questions should await those working in the ethics of assisted dying.

References

Bickenbach, Jerome. 1998. Disability and Life-Ending Decisions. In *Physician Assisted Suicide: Expanding the Debate*, ed. Margaret P. Battin, Rosamond Rhodes, and Anita Silvers. Routledge.

Braswell, Harold, and Rosemarie Garland-Thomson. 2023. When Anti-Discrimination Discriminates. *The American Journal of Bioethics* 23 (9): 35–38.

Colburn, Ben. 2022. Disability-Based Arguments Against Assisted Dying Laws. *Bioethics* 36 (6): 680–686.

Coleman, Diane. 2010. Assisted Suicide Laws Create Discriminatory Double Standard for Who Gets Suicide Prevention and Who Gets Suicide Assistance: Not Dead Yet Responds to Autonomy, Inc. *Disability and Health Journal* 3 (1): 39–50.

Della Croce, Yoann. 2020. Taking Oppression Seriously: Disability and Physician-Assisted Suicide. *Bioethica Forum* 13 (1): 22–27.

Gill, Carol J. 2010. No, We Don't Think Our Doctors Are Out to Get Us: Responding to the Straw Man Distortions of Disability Rights Arguments Against Assisted Suicide. *Disability and Health Journal* 3 (1): 31–38.

Hatherley, Joshua J. 2019. Is the Exclusion of Psychiatric Patients from Access to Physician-Assisted Suicide Discriminatory? *Journal of Medical Ethics* 45:817–820.

Lippert-Rasmussen, Kasper. 2014. *Born Free and Equal? A Philosophical Inquiry into the Nature of Discrimination*. Oxford University Press.

Montagna, Giacomo, Christoph Junker, Constanze Elfgen, Andres R. Schneeberger, and Uwe Güth. 2023. Long-Term Development of Assisted Suicide in Switzerland: Analysis of a 20-Year Experience (1999–2018). *Swiss Medical Weekly* 153 (3): 40010–40010.

Moreau, Sophia. 2010. What is Discrimination? *Philosophy & Public Affairs* 38 (2): 143–179.

Riddle, Christopher A. 2017. Assisted Dying & Disability. *Bioethics* 31 (6): 484–489.

Rooney, William, Udo Schuklenk, and Suzanne van de Vathorst. 2018. Are Concerns About Irremediableness, Vulnerability, or Competence Sufficient to Justify Excluding All Psychiatric Patients from Medical Aid in Dying? *Health Care Analysis* 26:326–343.

Schuklenk, Udo, and Suzanne van der Vathrost. 2015. Treatment-Resistant Major Depressive Disorder and Assisted Dying. *Journal of Medical Ethics* 41:577–583.

Shavelson, Lonny, Thaddeus M. Pope, Margaret P. Battin, Alicia Ouellette, and Benzi Kluger. 2023. Neurologic Diseases and Medical Aid in Dying Laws Create an Underclass of Patients Based on Disability. *The American Journal of Bioethics* 23 (9): 5–15.

Silvers, Anita. 1998. Disability Discrimination and the Duty to Protect Otherwise Vulnerable Groups. In *Physician Assisted Suicide: Expanding the Debate*, ed. Margaret P. Battin, Rosamond Rhodes, and Anita Silvers. Routledge.

Sleeman, Katherine, and Iain Chalmers. 2019. Assisted Dying: Restricting Access to People with Fewer than Six Months to Live is Discriminatory. *BMJ* 367:I6093.

Sneddon, Andrew. 2006. Equality, Justice, and Paternalism: Recentreing Debate About Physician-Assisted Suicide. *Journal of Applied Philosophy* 23 (4): 387–404.

Varelius, Jukka. 2015. On the Moral Acceptability of Physician-Assisted Dying for Non-Autonomous Psychiatric Patients. *Bioethics* 30 (4): 227–233.

Wright, Megan S. 2023. Current Medical Aid-in-Dying Laws Discriminate Against Individuals with Disabilities. *The American Journal of Bioethics* 23 (9): 33–35.

Chapter 2
Discrimination and Assisted Dying

Abstract This first section starts off by clarifying how the terminology of assisted dying and discrimination will be used throughout the book, drawing particular attention on the analytical account of discrimination developed by Kasper Lippert-Rasmussen. The key idea behind this section is to determine whether or not specific instances of inequality of access to assisted dying are actually discriminatory, as they are usually assumed to be in the literature. Three cases where discrimination is usually brought up are reviewed, those of disability, mental disorders, and existential suffering. I then move on to two less evident social groups which I argue are in fact subject to discrimination in access, those of rural communities and low-income groups. A mapping of discrimination in assisted dying is then provided to summarize the conclusions from the analysis. Such mapping is then refined to highlight the conceptual and normative differences between the different kinds of discrimination seen throughout the chapter.

Keywords Discrimination · Assisted dying · Equality · Access · Disability · Mental illness · Rurality · Social class

2.1 Preliminary Conceptual Remarks on Discrimination, Assisted Suicide, and Euthanasia

As it will appear clear soon enough, this short book puts great emphasis on the value of using conceptually sound and accurate terminology. As such, one will be quick to point out that this book title makes use of the term *assisted dying* instead of assisted suicide or euthanasia. This is deliberate, for this book's scope is larger than just the practices of assisted suicide or euthanasia, the chief difference between the two pertaining to the active role of the person whose life ends: in the former the requestor performs the killing act themselves with the help of a third party, in the latter it is a third party that performs the killing act for the requestor. While it is true that some jurisdictions have allowed one but not the other and that each of these practices come with their own set of ethical dilemmas, this book's ambition requires

Y. Della Croce, *Assisted Dying, Discrimination, and Structural Injustice*, SpringerBriefs in Ethics, https://doi.org/10.1007/978-3-032-11093-0_2

not to be bound by only assisted suicide or euthanasia. As it will become clear soon enough, the distinction between the two does not substantially matter for the task at hand here. There is one notable exception that we will soon address where this distinction matters, which occurs in jurisdictions where only assisted suicide is lawful and when the requestor is a person with a severe disability that prevents them from performing the killing act themselves. In this case, assisted suicide is deemed discriminatory and it is thus argued that the only way to have an equal assisted dying policy is to allow for the legalization of euthanasia. Aside from this specific case, this book's subject is that of *assisted dying* at large, understood as any form of practice where one requestor demands assistance to end their lives, whether they perform the final act themselves or a third party does it for them. It thus seeks to encompass both assisted suicide and euthanasia at the same time and the arguments made over its course are designed to be valid for any jurisdiction where one form of assisted dying or another is lawful, regardless of the law's precise content. In addition to granting us some conceptual latitude, proceeding in this fashion also allows for a non-parochial argument about assisted dying that is not tied to any particular legal framework. As we saw earlier the jurisdictions which have legalized some forms of assisted dying are many and growing, and the content of this book aims to be relevant in any one of them. As such, I will use the terminology of assisted dying through this book, understood in a sense that encompasses both assisted suicide and euthanasia.

Before diving into the analysis of specific instances of alleged discrimination in assisted dying, a few conceptual and analytical clarifications regarding the concept are in order. Chiefly, when one seeks to talk about discrimination, one must naturally start with an account of discrimination. This is not as straightforward a task as it first appears to be. Discrimination is indeed a widely, arguably overused, concept that deeply permeates public discourse and has long suffered from conceptual vagueness. As Vandenhole indeed stated in his review of international treaties that attempted to outlaw discrimination, there appears to simply be no universally accepted definition of discrimination (Vandenhole 2005). When one talks of discrimination, one may refer to something negative or something positive, think for instance of Jim Crow laws or affirmative action policies, the latter usually referred to as positive discrimination. One may forego the language of normativity at all and argue that discrimination simply refers to differential treatment based on any given characteristic: in this sense, any sort of club that enforces some kind of membership rule is discriminatory towards non-members. Discrimination may also be direct or indirect: perhaps the discriminatory effect suffered by some social group was not intended by the discriminator, be it through reduced opportunities or disparate impact. Many of these conceptual blurs have led discrimination to be more and more of a rhetorical tool instead of a proper philosophical framework, and such deviation into rhetorical grounds is visible in the context of assisted dying as well, as we have seen above and will see more clearly over the course of the next chapter. While the literature on discrimination is rich, it is unfortunately beyond the scope of this book to comprehensively attempt to address the many nuances and controversies that have become constitutive of the concept, and as such what we need for the present endeavor is simply an analytical account of discrimination that we can readily use to disentangle its many uses in the context of assisted dying.[1] To this end, Lippert-Rasmussen's 2014 *Born Free and Equal?* provides an extensive investigation into the analytical and conceptual nature of discrimination, specifically group discrimination, that proves ideal for the task at hand.

Lippert-Rasmussen starts off by highlighting what he refers to as the generic sense of discrimination, which broadly speaking consists of treating one worse according to some property that is either possessed or believed to be possessed by some other individual (Lippert-Rasmussen 2014: 15). More precisely, Lippert-Rasmussen synthesizes this generic account in the following way:

(i) There is a property, P, such that Y has P or X believes that Y has P, and Z does not have P or X believes that Z does not have P,
(ii) X treats Y worse than he treats or would treat Z by Φ-ing, and
(iii) It is because (X believes that) Y has P and (X believes that) Z does not have P that X treats Y worse than Z by Φ-ing. (ibid: 15).

Such a definition is unsatisfactory, for it appears overly broad and fails to provide any semblance of normative guidance. But let us first step away from ice-cold analytical definitions and look at some illustrations to better understand what is wrong with the generic account. Imagine you are casting a movie that is set in Korea and for which you seek to hire ethnically Korean performers. Now imagine I, a Caucasian, apply for the role. As you are looking to hire ethnically Korean performers, you understandably decide not to invite me for an audition, because I am Caucasian and thus, well, not ethnically Korean. In this instance, you have reduced my opportunity to star in your movie solely based on the fact that I hold the property to be Caucasian, which logically excludes the possibility for me to hold the property of being ethnically Korean. In the aforementioned generic sense of discrimination, such an instance would qualify as discrimination, though it is highly unclear what is morally dubious about it, if anything actually is. Similar cases can be imagined in the case of assisted dying: imagine you are my general practitioner, I rush to your practice completely drunk and demand that you allow me to access assisted dying because my lover left me, and I simply cannot bear the heartbreak. You decide to dismiss me because I am very clearly intoxicated, thus depriving me of a fair assessment of my request, let alone my opportunity to access assisted dying. Your decision to dismiss me is therefore based on my holding, at the time, of the property of being intoxicated, which leads you to treat me worse, in the sense of you reducing my opportunities, than if I had been sober. Again, according to the generic account of discrimination, this instance would qualify as discrimination as well. It appears evident that none of these examples seem to capture what we have in mind when talking of discrimination, at least not in a sense that is morally useful and meaningful: this generic account is just too broad. We thus need a more refined account of the concept to fit our purposes.

[1] See (Altman 2020) for an extensive survey of the philosophical and legal discussions on discrimination.

Lippert-Rasmussen does so by emphasizing on and refining two elements that are so far glossed over in the generic account of discrimination: that of the nature of the discriminatory act and that of the social standing and group identity of the discriminated, both of which combine to draw an account of *group discrimination*. With regard to the former, Lippert-Rasmussen highlights the need for the discriminatory act (that is, the Φ-ing) to be *relevant*, meaning the act itself must present some sort of morally dubious feature. Looking back at our previous example of an alternative version of myself where I pursue acting instead of research, and where I am still Caucasian, and I apply for a role designed specifically for people of Korean ethnicity, the fact that you decide to turn me down based on my ethnicity does not seem all that controversial. It would make no sense (and arguably would even be morally problematic) to hire me for the part instead of a Korean performer. While I may be discriminated against according to the generic sense of discrimination in this instance, it is difficult to see exactly why your allegedly discriminatory act is truly morally relevant, for it was reasonably expected of you to proceed as you did. In order for this example to satisfy the relevance condition, some tweaks would be required. You would, for instance, have to be making a movie for which you're just looking for a male lead, and we happen to live in a different world where Caucasians are systematically given less consideration and opportunities than others. In this alternate world, should I apply for the part of the male lead, and you decide not to invite me for an audition solely based on the fact of my ethnicity, then your action would be morally relevant. In other words, relevant discriminatory acts are harmful (ibid: 28).[2] Similarly, recall the drunken version of me storming into your practice to ask for assisted dying: it seems even more evident that you are not harming me by denying my request, and here again you would most likely be considered to be in the wrong for not acting in the way you did.

In addition to this relevance condition, which helps us in understanding *what* kind of act should matter when thinking about discrimination, Lippert-Rasmussen also puts great emphasis with regard to *who* is allegedly discriminated against. Discrimination is considered to occur when the alleged person discriminated against is part of what he refers to as a *socially salient* group. Lippert-Rasmussen defines a group as socially salient if the perceived membership of it is important to the structure of social interactions across a wide range of social contexts (ibid: 30). The key idea behind social salience is that not all groups, or more precisely group memberships, are relevant to social life. To illustrate, Lippert-Rasmussen highlights the difference between one's eyes color and one's race or gender: while it is clear that the latter does indeed affect one's social interactions in a variety of contexts, the same cannot be said of the former.[3] Some features that can indeed be used to classify

---

[2] While the relevance condition does make intuitive sense, Lippert-Rasmussen himself highlights that determining exactly what makes an act-type *relevant* needs more exploration and is not without controversy (see Lippert-Rasmussen 2014: 28, footnote 43). However, although not completely analytically clear-cut, this definition suffices for our here purpose of investigating assisted dying.

[3] Note that social salience is not fixed by any means, as such any group that may not be socially salient today could be in the future, and vice-versa.

people into distinct groups (i.e. having blue eyes) can be largely insignificant to one's social interactions (ibid: 31). Looking back at our two previous illustrations and applying the condition of social salience, we quickly see that the case of me not receiving an invitation for the Korean movie casting based on my ethnicity does in fact satisfy the condition: social salience does not require one's membership to *negatively* affect one's social interactions in a wide variety of contexts. As we saw earlier, the problem in this case lies with the relevance condition. As for the second case, a more interesting puzzle arises, though it is beyond the scope of this book to solve it. It appears evident that being drunk cannot, by definition, be socially salient: one cannot hold membership to a group defined by a transitory altered state. The same cannot, however, be said of alcoholics, to whom membership does indeed greatly affect one's social interactions. Nevertheless, as our example does not specify if this alternate version of me is an alcoholic, it fails as such to satisfy the social salience condition. Should we slightly change our premises to me being drunk to me being an alcoholic (but not drunk at the time of my appointment), as known by my general practitioner, and being denied access to assisted dying based on that fact, then the whole story would change: I would be part of a socially salient group and the relevance condition would also be satisfied, for it appears much less clear why I was denied access.[4]

These two necessary conditions being established, we can now lay out Lippert-Rasmussen's refined and complete account of group discrimination:

> X discriminates against Y in relation to Z by Φ-ing if, and only if:
>
> (i) there is a property, P, such that (X believes that) Y has P and (X believes that) Z does not have P,
> (ii) X treats Y worse than Z by Φ-ing,
> (iii) it is because (X believes that) Y has P and (X believes that) Z does not have P that X treats Y worse than Z by Φ-ing,
> (iv) P is the property of being a member of a certain socially salient group (to which Z does not belong), and
> (v) Φ-ing is a relevant type of act etc., and there are many acts etc. of this type, and this fact makes people with P (or some subgroup of these people) worse off relative to others, or Φ-ing is a relevant type of act etc., and many acts etc. of this type would make people with P worse off relative to others, or X's Φ-ing is motivated by animosity towards individuals with P or by the belief that individuals who have P are inferior or ought not to intermingle with others. (ibid: 45–46)

Crucial to highlight is the fact that this account of discrimination is not *moralized*, that is, it does not follow that if any situation qualifies as group discrimination according to the above definition, then this situation is necessarily morally wrong. Recall that Lippert-Rasmussen's analytical account of discrimination aims to clarify discrimination both conceptually and normatively. This last point is vital to the analysis this book seeks to undertake, for we want not only to be able to tell which

[4] This is of course not to imply that access denial in this situation is necessarily wrong, for one can imagine a myriad of good reasons to bar access to which being an alcoholic would be a good proxy. I am simply stating that the situation is less clear-cut than if one were to rush requesting assisted dying in a drunken state.

instance of unequal access actually is discriminatory, but also to distinguish between cases where such discrimination is wrongful and where it is not. In the case of assisted dying, it is indeed perfectly possible to imagine cases of discriminatory access that satisfy the above definition that are simultaneously not morally problematic, or even morally desirable. But we must leave these points aside for now, for we will get back to them shortly.

There is one last, but not least, idea we need to address before moving on to the analysis properly speaking, that is the idea of *indirect* discrimination. The above account of group discrimination encompasses cases where an agent is biased towards a certain other agent based on their membership (or perceived membership) in a socially salient group and treats them relatively worse because of the property of being perceived to be a part of this group. But what about cases where an act does not, in any way, reflect bias on the part of the discriminator? While the former can be said to constitute direct discrimination, the latter is often considered as indirect discrimination. The idea of indirect discrimination is essential to an analysis of the usages of discrimination in the ethics of assisted dying, for it is at the foundation of some of the core arguments against legalization. Recall the fundamental idea behind the slippery slope argument: legalizing assisted dying will necessarily lead to the unintended hastened deaths of the most vulnerable, and especially of those with disabilities. The key word here is unintended, meaning that it is assumed that legislators or those in favor of assisted dying have no desire to *actually* kill, or foster the killing, of those with disabilities: killing those with disabilities will only be the inevitable consequence of adopting laws that permit assisted dying. This argument rests on the idea of the potential *disparate impact* that legalization may have on the most vulnerable, which has often been said to be discriminatory. This is typically what is thought of when speaking of indirect discrimination.

Indirect discrimination has a contested history in philosophical scholarship, being sometimes labelled as a legal construct that serves only judicial purposes with little philosophical substance (Eidelson 2015: 19).[5] There are nevertheless strong intuitions that indirect discrimination is an important concept in today's political, social, and legal landscape. It indeed seems like constant efforts have been made to get rid of discriminatory laws and systematically punish discriminatory acts, yet stating that discrimination has been eradicated sounds like a fantasy. There must be more to the idea of discrimination than the direct, intentional account we have sketched out above. What, for instance, of cases where prejudice is so deeply anchored within a culture that it leads to discriminatory conduct without even realizing that it is discriminatory, let alone intending for it to be? Or what about cases where pure indifference leads to discriminatory states of affairs? Lippert-Rasmussen here again provides us with useful guidance in sketching out an analytical account of indirect discrimination. As we highlighted, the key feature of indirect

[5] Here again, an extensive discussion regarding the validity of indirect discrimination as a philosophical concept is beyond the scope of this short book. However, see (Cossette-Lefebvre 2020) for an in-depth exposition of Eidelson's criticism as well as arguments in favor of a normative idea of indirect discrimination.

discrimination is the idea of a disproportionate burden that is inflicted on a group of people, which does not involve any form of bias or intentionality on the part of the agent that has brought about such a burden (Lippert-Rasmussen 2014: 55). Incorporating these elements into his account of direct discrimination and modifying it accordingly, Lippert-Rasmussen reaches the following analytical account of indirect discrimination:

> A policy or act is indirectly discriminatory against (members) of a certain group, G, if, and only if:
>
> (i) It reflects no bias on the part of the discriminator against members of G on account of their being members of G, and it is not the case that the relevant policy or act reflects a bias, or unbiased attitude, of the discriminator in favor of members of G qua members of G.
> (ii) It relevantly disadvantages, i.e., on average or in the case of most members of the relevant socially salient group, locally or globally, relevant to other relevant, socially salient groups.
> (iii) G is a socially salient group.
> (iv) The disadvantages referred to in (ii) would not have occurred in the absence of past or present direct discrimination by the agent of the policy or act in question against G. (ibid: 72).

Conditions (i), (ii), and (iii) are relatively straightforward if we keep in mind what the core features of direct discrimination were and what the core features of indirect discrimination should be. Condition (iv), which Lippert-Rasmussen refers to as the *causal* condition, emphasizes the fact that indirect discrimination cannot exist in a vacuum: it must place itself within a logic of perpetuated disadvantage for it to qualify as discrimination. This means the existence of indirect discrimination is contingent upon the existence of, or the past existence of, direct discrimination regarding the experience of a given socially salient group. There is, thus, to an extent, a structural component to the idea indirect discrimination: harm would not have occurred with pre-existing logics of exclusion and disadvantage that fall outside the realm of an agent's intention to do harm, and as such must find at least part of the origins within social structures, and possibly structural injustices. We are now almost fully equipped to proceed with our conceptual analysis, armed with our analytical accounts of direct and indirect discrimination: either each of the cases we will scrutinize falls within one of these categories, or they will be deemed not to qualify as discrimination. This will suffice in distinguishing between cases where the use of discrimination is substantial as opposed to rhetorical, and as we shall see will also shed light on some forms of inequality of access that have not been qualified as discriminatory but may actually well be. Once this conceptual mapping is established, we can then move on to a normative assessment of discrimination in assisted dying, particularly insofar as the interplay between structural injustice and unequal access is concerned.

2.2 The Many Faces of Discrimination in Assisted Dying

This section is divided into two subsections, each with a distinct methodological approach and unified by the common thread of inequality of access. The first of these is dedicated to cases for which restricted access has been described as discriminatory in the literature: disability, mental disorders, and existential suffering. My goal here is to start from what is considered discriminatory in these cases and work our way back to our analytical account of discrimination to evaluate whether they can actually be accurately described as discriminatory or not. The second subsection takes a different approach: I take a look at two cases that are *not* usually qualified as discriminatory (and where there is clear inequality of access) and evaluate whether they may count as discrimination or not. These cases are those of rural communities and low-income groups. What I intend to do through this section is to show that our understanding of discrimination in assisted suicide may be flawed, for cases we usually think of as discriminatory may not be and conversely, cases that we usually would not think of as discriminatory may actually well be.

2.2.1 *Disability*

The conceptual use of discrimination with regards to disability can be split between two categories: first those that use discrimination towards people with disabilities as a mean to argue about the permissibility of assisted dying and second those that use discrimination to criticize the boundaries of existing assisted dying legislations. The first category is thus naturally split between two further sub-categories, those that are pro-legalization and anti-legalization. This variety of viewpoints is largely due not only to the complexity of the ethical matter at hand but also to the crucial fact that there exists no consensus on the question of assisted dying within disability rights organizations or the scholarship from disability studies, echoing the wide breadth of opinion that is found amongst non-disabled citizens (Colburn 2022).

The idea behind the arguments of those that have used discrimination as a mean to argue against legalization has already been alluded to in the introductory section of the book and somewhat follows the seminal slippery-slope concerns first brought to the fore by Yale Kamisar in 1958: legalizing assisted dying may be potentially dangerous inasmuch as some socially-oppressed individuals may be coerced to kill themselves due to their existence being "inconvenient" or socially "unfit" (Kamisar 1958).[6] Not Dead Yet, a U.S. based anti-assisted dying advocacy group, has been

[6] In slippery-slope arguments, the category of the socially vulnerable often includes the disabled as well as the elderly. The argument is substantially similar in either case, there has however been less direct use of discrimination towards the elderly as opposed to the disabled and thus I do not dedicate a special subsection to this case. One could however reasonably assume that the conclusions, as far as the analytical account of definition is concerned, are similar as well.

amongst the most vocal as far as the alleged discriminatory nature of the legalization of assisted suicide towards the disabled is concerned. In an Amici Curiae Brief addressed to the U.S. Supreme Court in the 1997 landmark case *Vacco v. Quill*. The brief warns of the legalization of assisted dying being the most lethal form of discrimination against people with disabilities, inasmuch as it is the "ultimate expression of society's fear and revulsion regarding disability".[7] While Not Dead Yet's stance may be extreme and dubious with regards to its paternalistic overtones (Bickenbach 1998), the substance of their argument is still endorsed by some. One of the feared consequences of legalization is the creation of a discriminatory double-standard on who gets suicide prevention and who gets suicide assistance, where people with disabilities will be overwhelmingly pushed towards the latter (Coleman 2010; Golden and Zoanni 2010). The argument is somewhat easy to summarize: since people with disabilities live in an ableist world where their lives are systematically devalued, then the permissibility of assisted dying will either create perverse institutional incentives to point them in the direction of assisted death where other solutions would have been more adequate (Batavia 1997), or will convince them further that their lives are not worth living and nudge them towards assisted death, since their existence is already marked by systemic oppression and discrimination and subsequent inequality of autonomy. Legalizing assisted dying is thus discriminatory towards the disabled because of the disparate impact of legalization.[8]

Discrimination has also been employed as a means to argue in favor of the legalization of assisted dying. While the anti-legalization camp largely rested its case on the basis of inequality of autonomy and unequal concern, the pro camp's argument is rather concerned with equality of opportunity. The argument goes as such: people with disabilities, depending on the nature of their disability, may face obstacles in committing suicide that non-disabled people will not. If suicide is legally permitted, then it is necessary for people with disabilities to be able to exercise their implicit right to suicide. Equality demands this right to be not merely formal but also substantial and as such, should the disabled person not be able to end their life on their own, they are entitled to suicide assistance on the grounds of equality (Sneddon 2006).

Such claims to equality have in fact been made before, most notably in the *Rodriguez v. British Columbia* case brought to the Canadian Supreme Court in 1993. Sue Rodriguez, a woman in her early forties diagnosed with amyotrophic lateral sclerosis, challenged the blanket prohibition on assisted suicide found in Section 241(b) of the Canadian Criminal Code (Rodriguez v. British Columbia 1993). Her condition was degenerative and terminal, and while she remained

---

[7] This view is closely related to what is sometimes referred to as the expressivist objection to the legalization of aid-in-dying. The expressivist objection states that legalizing assisted suicide and/or euthanasia sends a signal that the lives of the vulnerable and oppressed are not worth living and thus, as a matter of equal concern, these practices should not be legalized. For an in-depth review and rebuttal of this view, see (Braun 2023).

[8] This view has notably been strongly criticized because of its overtly paternalistic and victimizing portrayal of people with disabilities, both from within the sphere of disability studies and outside (Silvers 1998; Riddle 2017; Della Croce 2020).

mentally competent, she anticipated that she would soon lose the physical capacity to end her life unaided. Rodriguez sought the legal right to receive assistance in dying at a time of her choosing, before complete physical deterioration made it impossible. Rodriguez's legal challenge was grounded in the Canadian Charter of Rights and Freedoms. She argued that the prohibition violated her rights to life, liberty, and security of the person (Section 7), as well as her right to equality (Section 15), by denying her the ability to make decisions concerning her own body and death. Crucially, she claimed that the law discriminated against people with disabilities by allowing able-bodied individuals to end their lives without legal interference while denying that same possibility to those who, due to physical impairment, required assistance. In a narrow 5–4 decision, the Court upheld the prohibition. The majority acknowledged that Rodriguez's Section 7 rights were engaged but concluded that the infringement was justified in light of the state's interest in protecting vulnerable persons and upholding the sanctity of life. The equality claim was also rejected, on the grounds that the law applied uniformly to all individuals regardless of ability status. The dissenting justices, however, noted that this formal equality failed to account for the law's unequal effects: only those with severe physical impairments were effectively barred from acting on a wish to die. While unsuccessful, Rodriguez's case played a significant role in shaping the legal and ethical discourse around assisted dying and disability. It exposed how a facially neutral law could disproportionately burden individuals with physical disabilities and raised questions about whether a right can be meaningfully exercised if it is inaccessible to certain groups without accommodation. The issues raised in *Rodriguez* would eventually be revisited in *Carter v. Canada* (2015), where the Supreme Court overturned the precedent and acknowledged that a blanket ban on assisted dying could produce unconstitutional and discriminatory effects.

Legalizing at least a minimal form of suicide assistance is thus necessary in order to avoid discriminating against the disabled and guarantee substantial equality of rights as far as self-determination is concerned. While this line of argument pertains to the dichotomic approach, some of its core features have also been used in arguing against the discriminatory aspects of existing assisted dying laws. Indeed, while assisted dying legislations have been adopted in a variety of jurisdictions, the central problem of substantial capacity for suicide has all but disappeared. Assisted suicide laws retain a self-administration requirement, which can prove problematic, especially so with requestors who suffer from neuromuscular disabilities such as amyotrophic lateral sclerosis (Shavelson et al. 2023). In effect, despite legislation having been adopted, the problem remains essentially similar: those who do not have the physical capacity to effectively end their lives are excluded from accessing assistance. Moreover, it has been argued that the self-administration requirement is not the only way in which current laws discriminate against people with disabilities: the potentially narrow understanding of decisional capacity and requirement for one-to-one discussion without the presence of others between the requestor and the physician may also hinder access for people with cognitive and communication disorders (Wright 2021, 2023).

How do these various claims of discrimination fare when looked at through the lens of the theoretical framework of direct and indirect discrimination fleshed out above? It is evident that the group of people with disabilities qualifies as a socially salient group and that the existence, non-existence or form of assisted dying legislation relevantly disadvantages the group relatively to other social groups. This is true for all of the aforementioned instances. In the case of those who use discrimination as a mean to argue against legalization, there is however room for doubt as to whether the discrimination they put forward belongs to the direct or indirect kind. Recall that for a policy to qualify as indirect discrimination it must reflect no bias on the part of the discriminator towards the concerned group. The arguments at stake appear to be in somewhat of a disagreement with respect to this very point. Following Not Dead Yet's line of reasoning, one could reasonably assume that the potential threats associated with legalization are not mere oversights; they signal a clear disdain for the lives of the disabled and as such the negative consequences of legalization may be said to be deliberate. This however needs to be nuanced. Carol Gill has for instance insisted that the views of the anti-legalization camp are not to be equated with a "doctors are out to get us" perspective (Gill 2010), instead insisting on the perverse unintended effects of legalization. This appears more in line with the use of the disparate impact terminology, often present in this literature and aligned with our definition of indirect discrimination. As this manuscript's focus is on the philosophical literature and not on public and associative discourse, it is thus reasonable to reject Not Dead Yet's radical view in favor of the philosophically sounder view of anti-legalization scholars. As for the final requirement, it is also trivially satisfied as it is evident that people with disabilities have been the victims of direct discrimination in the past on the part of both policy and lawmakers in a variety of contexts. The conclusions we reach are identical when scrutinizing parity arguments. In these latter cases, there is however no doubt that there is no bias on the part of the discriminator, placing them firmly in the category of indirect discrimination.

2.2.2 *Mental Disorders*

The inclusion of mental disorder, particularly psychiatric disorder, as a valid criterion of eligibility to assisted dying has been the subject of many controversies, occupying without doubt the largest portion of debates on the continuum approach. The population targeted by the legalization of assisted dying has after all mainly consisted of those that suffered from incurable diseases with poor prognosis which caused unbearable pain; most of the public opinion that supports legalization does it on the basis of these requirements (Schuklenk et al. 2011). In addition, the capacity for autonomous decision-making has traditionally been considered to be a necessary safeguard in ensuring the practice of aid in dying does not bring about some of the undesired negative consequences brought to the fore by proponents of slippery slope arguments. These requirements are still reflected in a significant number of contemporary jurisdictions, although some have started to stray away from them in a bid to craft more inclusive legislation that guarantees a wider array of requestors to be allowed access.

The above-mentioned criteria for eligibility prove indeed highly problematic for those suffering from mental illness who wish to request assisted dying. Psychiatric disorders cannot be said to have poor prognoses unlike cancer or degenerative diseases. The requestor's capacity for autonomous decision-making may also be hindered by symptoms of the disease, most typically in the case of depressive disorders (Ogilvie and Potts 1994; Tsou 2013; Pronk et al. 2019). Access to assisted dying for psychiatrically ill patients has thus been barred in a vast majority of jurisdictions.

Some have however questioned this restriction of access on the basis of its alleged discriminatory nature. While it is true that some patients may suffer from limited autonomous decision-making capacity, this does not imply that *all* mentally disordered requestors, including those suffering from treatment-resistant major depressive disorder, are necessarily incompetent (Schuklenk and van der Vathrost 2015; Rooney et al. 2018). This point was also reaffirmed by Hatherley who insisted on the divide between mental and physical illness to be weaker than commonly assumed, and certainly not strong enough to warrant exclusion from assisted dying for the mentally ill (Hatherley 2019). Varelius further criticized the competency requirement as inconsistent with the current practice of passive euthanasia and as such should not be relied on to justify exclusionary laws (Varelius 2015). A related but distinct form of this latter argument posits that the suffering from physical illness is the only legitimate reason that ought to be accepted in requesting assisted suicide. This might either be the normative conclusion of the expertise-based, or epistemic, argument or a distinct line of reasoning. "Legitimate" might refer to the procedural rules that justify or not the existence of a practice, in this case the legitimacy argument is tied to the epistemic argument. As just mentioned, in this sense the legitimacy argument is in fact not an argument in itself but merely the conclusion of the epistemic argument. There is however another sense in which the use of legitimacy does not refer to procedural rules but rather to the value of the request itself. Legitimacy in this sense calls to the normative validity of suffering, stating that the only suffering that is *worth* considering when thinking about assisted suicide is of the physical kind. In sum, exclusion of the mentally ill from assisted suicide may be justified in terms of autonomy, epistemic limitation, or legitimacy of request (respectively the autonomy-based, epistemic-based and legitimacy-based arguments).

Unlike the aforementioned disability case, the discrimination at play here cannot be said to be of the indirect kind. The fact that the mentally ill cannot access assisted suicide is because they are explicitly banned to do so; it is not an oversight or unforeseen consequence of legislation. Recall that for a policy to be considered as direct discrimination, it must be the case that the discriminator believes that the socially salient group in question possesses some property P and that it is because of this property that the group is treated relatively worse than other groups. I assume it to be uncontroversial to consider the mentally ill as a socially salient group, particularly when acknowledging the stigma associated with mental illness and psychiatric disorders (Hinshaw and Stier 2008). It has been counter-argued that the exclusion of the mentally ill from assisted suicide does not necessarily constitute discrimination for it rather reflects their inability to fulfill some eligibility criteria, which may themselves be ethically justified or not (Braun et al. 2024). The authors point that the exclusion of the mentally ill from access is due to how assisted dying itself is understood and what key ethical principles it ought to foster. While this point is valid, this does not change the fact that current laws which exclude mental illness as a valid motivation from assisted suicide due to the alleged lack of competency from requestors still constitutes discrimination. It is not far-fetched to assume that the erroneous perception of the mentally ill's lack of capacity for autonomous decision-making is somewhat related to stigma and subsequent perception bias of both clinicians and lawmakers. The mentally ill are thus being treated worse in virtue of being mentally ill by exclusionary legislation, thus being targets of direct discrimination, at least in the current state of legislations that forbid them access. It is worth highlighting that in this short discussion I have mostly engaged with autonomy-based arguments in favor of exclusion. Those are the most benevolent, as opposed to those arguments that reject the legitimacy of mental suffering. However, the line of thought remains the same for both the epistemic as well as legitimacy-based arguments in favor of exclusion: the group of the mentally ill is still as socially salient, is still worse off,[9] and is still treated differently on the basis of some property that is characteristic of the group. Psychiatric disorder is always in one form of another the culprit and cannot be dissociated from the group of the "mentally ill": in the epistemic-based argument its impact on the requestor's life is said to be beyond the assessor's comprehension and in the legitimacy-based argument it is said to be a sort of second-class suffering that does not warrant suicide assistance.

The autonomy-based argument for the exclusion of the psychiatrically ill may seem to echo some of the arguments put forward in the previous section on disability. They however differ with respect to two distinct points. First, the normative endpoint of autonomy-based arguments differs in the case of disability: while in the case of mental illness autonomy is used as a metric to limit access to those for whom it is lacking, in the case of disability it is used as a mean to argue for a ban on assisted suicide precisely because of the seemingly lack of autonomy on the part of the vulnerable (in this case, people living with disabilities). This echoes the continuum/dichotomic distinction we have fleshed out above. Second, the source of the alleged deficit in autonomy greatly differs between the mentally ill and the disabled: for the former autonomy is deemed to be lacking because of their illness whereas for the latter it's the social conditions that shape the experience of disability that is the culprit. There is however one significant exception to this distinction which is the case of cognitive disorders and disabilities. This category includes all forms of disorders and disabilities that may effectively hinder the capacity for autonomous choice but are distinct from psychiatric disorders insofar as the source of the

[9] As mentioned earlier, the aim of this section is to conceptually, and not normatively, assess claims of discrimination. It can be argued that autonomy-based arguments are justified in virtue of making the group actually better off, making discrimination actually morally justifiable and even desirable.

affection is considered.[10] This includes for instance developmental disorders as well as degenerative disorders such as dementia or Alzheimer's. While the heading of mental disorder is useful in order to build up a general and inclusive categorization that can conceptually incorporate such a variety of cases, it remains crucial not to amalgamate them into one single cohesive group, especially as far as the present inquiry about discrimination is concerned. It would be mistaken to say for instance that two individuals respectively suffering from schizophrenia and Alzheimer's belong to the same socially salient group. Nevertheless, while it is true that their respective afflictions substantially differ in how they engender membership to socially relevant groups and in how they structure social interactions, it also is true that they do structure social interactions and as such provide a sufficient basis to be each qualified as a socially salient group. They are grouped together in this subsection not in virtue of their shared membership in one single group but instead in virtue of the scope of the autonomy-based argument in favor of exclusion. This does not exclude the possibility that individuals with developmental disorders can also simultaneously belong to the aforementioned category of disability and thus be within the reach of the disparate impact argument we have seen before. There is nothing preventing any group from belonging to multiple categories at the same time; this will become especially relevant as we explore structural cases of exclusionary access later on. What truly matters for our inquiry is that the autonomy-based argument in favor of exclusion of access to those with mentally disorders, be them of psychiatric, developmental or degenerative nature, constitutes a case of direct discrimination against all groups that do not fit the cognitive criterion of eligibility to assisted suicide.[11]

---

[10] The conceptual connections and distinctions between terms like "mental", "cognitive", "illness", "disorder" and the likes are blurry and not yet philosophically settled. Note that even the use of "mental disorder" is conceptually as well as normatively contested; I intend to make no philosophical claims through the use of this terminology other than it being broad enough to incorporate psychiatric, developmental and degenerative disorders. Jennifer Radden and Jonathan Tsou's entry in the Stanford Encyclopedia of Philosophy provides a helpful and exhaustive summary of these conceptual disputes; see (Radden and Tsou 2024).

[11] There is another significant group that is targeted by the autonomy-based argument in order to restrict access to assisted suicide, that is children. The logic behind the exclusion of children from assisted suicide is in fact very similar to that of those with mental disorders since they are usually not considered competent and have to wait until their legal majority to make an eligible request. This has been criticized by some on the very same grounds that exclusionary access to those with mental disorders has been criticized: while some of the group's members may lack competence, this does not imply that it is an intrinsic feature of the group to systematically lack the capacity for autonomous decision-making. See (Chhikara 2017) for an example of this view applied to assisted suicide for children. Interestingly, it can also be said that children are actually facing direct discriminatory treatment with regards to access to assisted suicide since they fit our analytical account. At the time of writing, euthanasia for children is legal in Belgium (regardless of age) and in the Netherlands (for children aged above 12; scheduled to change to include all ages) (de Keijzer et al. 2023).

2.2.3 *Existential Suffering*

As far as eligibility criteria regarding the nature and magnitude of suffering are concerned, there is one other case beyond assisted suicide requestors with somatic and psychiatric illnesses: those that suffer from neither. Note that this does not imply that the requestor does not actually suffer from any sort of illness. An individual requesting assisted suicide on the basis of existential suffering may well be physically or mentally ill, the key difference being that their motivation for assisted suicide is not grounded in illness but on some other, potentially derivative, kind of suffering. These cases are referred to as pertaining to existential suffering where individuals are said to be "tired of life" and are thus not willing to continue living. It is crucial not to equate existential suffering with depressive disorder: in instances of existential suffering the desire to die is not deemed to be symptomatic of any underlying illness whatsoever. It might appear unsatisfactory to define the class of requestors who existentially suffer as those who do not suffer from any physical or mental illness. Providing a clear and exhaustive definition of existential suffering is however a more challenging task than what seems at first glance. Jukka Varelius attemps to define existential suffering as "(…) cases of unbearably distressed persons whose suffering is caused by their conviction that their life will never again be worth living" (Varelius 2014a, b). Recent empirical work has somewhat challenged this definition by identifying a plurality of meaning of existential suffering, including a variety of features such as physical decline, the feeling that life is over, loss of hope in the future, being a financial burden and loneliness amongst others (Gaignard and Hurst 2019). While it is crucial to acknowledge the multifaceted dimension of existential suffering, particularly insofar as clinical practice is concerned, Varelius's definition will suffice for the present analysis.

Requests for assisted suicide based solely on existential suffering are usually not admissible in legislations that allow for assisted dying. Judging from the exclusion of mentally ill individuals we have explored before, this may appear hardly surprising. Things are however not as straightforward. As seen before, the exclusion of the mentally ill is usually justified on the basis of a deficient, or complete lack of, capacity for autonomous decision-making due to the nature of the illness, be it because the wish to die is symptomatic of the illness or because the illness itself has allegedly hindered the requestor's autonomy. Since existential suffering is defined by the absence of illness or relevant illness-related features, none of these claims hold ground when arguing for the exclusion of existential suffering as a valid motive for requesting assisted death. The problem thus lies in the nature of existential suffering, or more particularly in the difficulty of its assessment on the part of physicians. Varelius has argued against this expertise-based argument on the ground that there is no truly objective way to evaluate the severity of physical pain either, the difference between physical and existential suffering consisting of mostly an epistemological fiction (Varelius 2014a, b).[12]

Unlike the cases of disability and mental illness, the exclusion of existential suffering as a valid criterion for requesting assisted suicide has not previously been qualified as discriminatory. As we have seen, some have argued that the exclusion is in fact unwarranted and morally objectionable, albeit not on discriminatory grounds. This makes intuitive sense: the class of individuals who request assisted suicide on the basis of existential suffering not only does not constitute a socially salient group and therefore exclusionary access cannot be made upon some P-property that is believed to be shared by members of the group, as there is no group in the first place. While of speculative nature, there does, however, exist an argument that could be brought up in an attempt to define the group of existential suffering-based requestors. It may be argued that existential suffering is not uniformly distributed across the population and is in fact a broad label that actually mostly captures the suffering of the elderly, and not that of young men who just experienced their first heartbreak. Such a view would of course need to be supported by empirical evidence, there however does seem to be some intuitive merit to it. It is nevertheless not enough to solve the aforementioned problem: even if the group of existential suffering-based requestors is actually a proxy for the elderly, which uncontroversially constitute a socially salient group, it remains unclear what property believed to be held by this group and not by other groups may be used to motivate discriminatory conduct. What is clear as of now is that the exclusion of existential suffering as valid criterion for access to assisted suicide does not constitute discrimination for the simple reason that it does not appear to discriminate against anyone in particular.

A similar problem arises for another case that is sometimes alleged to be discriminatory: those that suffer from diseases that will eventually kill them but whose prognoses extend beyond what assisted laws allow for eligibility. This typically includes progressive and degenerative diseases like dementia or amyotrophic lateral sclerosis such as in the case of Sue Rodriguez we have seen before. The idea is that those who suffer from this kind of disease will be the ones who suffer for longer and thus are those especially in need of the availability of assisted suicide (Sleeman and Chalmers 2019). Although there most certainly is a justice-based case to be built here, this kind of exclusion can hardly be qualified as discrimination for the same reasons we have seen when assessing existential suffering. There exists no such socially salient group as those who suffering from progressive diseases with prognoses that extend beyond the criteria for eligibility.[13]

[12] While there is no space to address the conceptual disputes regarding the nature of suffering throughout these pages, see (Edwards 2003) and (Nelson et al. 2025) for useful theoretical and normative work regarding the ambiguity of suffering and how the distinction between so-called "physical" and "existential" suffering may not be as clear-cut as it is made out to be in the practice of medicine.

[13] Although not quite exactly equivalent to the group at stake here, Philip Reed has recently argued that there exists such a thing as "terminalism" which amounts to discrimination against the dying (Reed 2023). In this sense, Reed aligns himself with critics of current eligibility criteria that

2.2.4 *Rural Communities*

We now turn to less obvious cases of inequality of access, at least as far as the attention of the canon of bioethics is concerned. Obviously, there are no laws that directly exclude assisted suicide requestors on the basis of spatial localization in any jurisdiction that allows for aid-in-dying. Nevertheless, this latter fact does not imply that access to all regardless of spatial localization is guaranteed. The ways in which access to aid in dying is hindered for rural communities naturally vary depending on the geographical contexts in which legislations exist. Access for rural communities may not be too problematic in small and densely populated countries like Belgium, the Netherlands or Switzerland, but can prove to be a challenge in larger geographical entities where the urban-rural gap is starker and where rural communities endure more isolation, as is for instance the case in Australia or Canada. This present analysis focuses on these two countries as clear instances of failed equality of access to assisted suicide for rural communities, albeit for different reasons. Let us take a closer look at both of these cases to flesh out what exactly is meant by what I claim here.

Let us start with Canada first. One key element to keep in mind when looking at unequal access to assisted suicide for rural communities is that the source of inequality of access is not to be found within the laws regulating aid-in-dying themselves, in other words those laws are not directly exclusionary nor are they intended to be, but rather arises from complex interactions between those laws and others that regulate other areas of the practice of healthcare. These interactions, while not necessarily problematic in themselves, end up creating burdens that disproportionately affect rural communities. A critical point to highlight is that the points that I address here pertain specifically to assisted dying. What I intend to mean by that is that inevitably there will be, and there are, challenges for rural communities to access aid-in-dying solely because of the fact that difficult access to aid-in-dying, a medical practice, mirrors the difficult access to healthcare in general that burdens rural communities (Hartley 2004). While this is of course morally problematic and plays somewhat of a role in the present analysis as a background element, this is not the focus of my argument. Instead, I wish to focus on regulations that specifically affect access to aid-in-dying but would not necessarily prove problematic in other areas of care delivery. As such, the two elements that I will address here are conscientious objection, either individual or institutional, and the prohibition of discussing suicide over

include only the terminally ill albeit for a very different reason: the fact that access be restricted to the terminally ill in his view does not constitute discrimination against all those who do not fit the criterion but rather against those who actually are terminally ill. This is because he defines the terminally ill as a socially salient group to whom medical options are wrongfully denied on the basis of their terminal illness, thus constituting discrimination. As aid-in-dying is only one of the instances in which he argues this kind of discrimination is displayed and not its main focus, I do not consider his views here. For a discussion and critique of the relevance of terminalism see (Han 2023; MacKenzie 2023; Riddle 2023; Sarbey 2023).

telehealth consultations. The former applies to both Canada and Australia, whereas the latter pertains to Australia only.

Perhaps the best way to introduce the issue at hand here is through a real-life event that occurred in 2017, shortly after Canada's adoption of Bill C-14 that legalized MAiD. After suffering from lung failure, Mr. Horst Saffarek decided to end his life through the use of MAiD. Mr. Saffarek was a resident of Comox, British Columbia, a small town in Canada's Western province where the only hospital is St Joseph's General, a catholic institution. St Joseph refused to provide MAiD on religious grounds, exercising their right to institutional conscientious objection. This refusal left Mr. Saffarek with the only option of transferring to the closest hospital that would perform the procedure, some 115 km away in Nanaimo, British Columbia. The transport to Nanaimo's facility brought excruciating pain to Mr. Saffarek and he passed away shortly after arrival, without having been able to access MAiD.[14]

The fact that access to assisted dying in Canada is challenging for rural communities is not new or surprising in any way (Schiller 2017), largely echoing the pre-existing difficult access to other healthcare services. However, following our intuition may lead us to start thinking that something about Mr. Saffarek's tragic final days is not just a consequence of pre-existing healthcare access disparities. It is true that MAiD is an ethically controversial practice for many and may, rightfully or not, be subjected to conscientious objection (Cowley 2016; Savulescu and Schuklenk 2017); a topic that is contentious not only amongst philosophers but also within public opinion and for stakeholders (Close et al. 2023; Knox and Wagg 2023). The case of Mr. Saffarek nevertheless appears somewhat extreme, and one may be quick to ask the following question: would a similar situation have unfolded in a metropolitan area such as Vancouver, Toronto, or Montreal? Reports from Canada clearly indicate that the answer here is a negative[15] and there is empirical work that has demonstrated the negative impact of such forced transfers on MAiD requestors (Wiebe et al. 2022). Was St Joseph's General right in exercising their right to institutional conscientious objection knowing they were the only reasonably accessible provider of such care for Mr. Saffarek? This latter question has sparked a significant amount of debate in bioethics and as of the time of writing there is no clear consensus on the matter (Shadd and Shadd 2019; Sumner 2019; Kirby 2021). While providing a definitive answer to this question is beyond the scope of the present analysis, we are nonetheless left with a striking observation: the right to conscientious objection, as far as MAiD is concerned, disproportionately affects rural residents due to the fact that rural communities are already deprived of options with regards to healthcare institutions in the first place. This calls back to my previous comment on the complexity of interactions in the introductory part of this subsection: neither of these statements is *sufficient* on its own to give rise to a moral

[14] https://www.cbc.ca/radio/thecurrent/the-current-for-january-11-2018-1.4481312/should-catholic-hospitals-have-to-provide-access-to-medically-assisted-dying-1.4482372 Accessed on May 16th 2024.

[15] https://www.cbc.ca/news/health/medically-assisted-dying-access-one-year-later-1.4165936 Accessed on May 16th 2024.

problem with regards to MAiD; they are both *necessary* and pertain specifically to assisted dying. As such, the troubling inequality of access to aid-in-dying endured by rural communities cannot be only explained by pre-existing disparities, but those disparities combined with the existence of conscientious objection laws create unique problems for rural residents who seek to access MAiD. There is furthermore another consideration to factor in that differs from other cases of healthcare services access disparities which is the extreme vulnerability that a vast majority of assisted dying requestors are subjected to regarding their capacity for freedom of movement. Using the terminology of theorists of spatial injustice, it can be safely argued that assisted dying requestors have low *motility*, defined as the capacity for mobility (Flamm and Kaufmann 2006). Taking a look at this literature is not only helpful as far as terminological clarity is concerned but also illuminates the spatial injustices that affect rural communities. Shedding the light on the spatial distribution of goods as a social product and not as a mere context in which distributive injustices happen (Pirie 1983) provides us with the tools to conceptualize the structural aspect of unequal access to assisted dying for rural communities, a point to which we will come back to later on when discussing whether such inequalities can accurately be qualified as discrimination or not.

We can now turn to the Australian case.[16] I will assume the reader to be already familiar with the sheer size of Australia and its subsequent strong urban-rural gap. About 28% of all Australians, which amounts to about 7 million citizens, are said to live in remote and rural areas of the country (Australian Institute of Health and Welfare 2020). Similar findings as to how conscientious objection to VAD disparately impacts rural communities have surfaced as well (White et al. 2023). The ethical considerations at stake here are similar to what I have just sketched out above and I will thus not return to them. There is however a unique feature of the Australian legal landscape that hinders access to rural communities in addition to the difficulties posed by conscientious objection and the lack of physicians in remote areas, that is the legal prohibition to discuss "suicide-related material" over carriage services. Carriage services are defined as services for carrying communications by means of guided and/or unguided electromagnetic energy.[17] Interestingly, these offences were introduced in 2005 to mitigate the negative impact of suicide chatrooms but also to undermine the promotion of suicide methods for the terminally ill, notably championed by Philip Nitschke (Del Villar et al. 2021). Philip Nitschke is a famous Australian physician actively engaged in right-to-die advocacy

[16] Note that as of November 2023, all states have implemented legislation that allows for aid-in-dying, referred to as the Voluntary Assisted Dying (VAD) Act. This is however not the case in the Northern Territory as well as in the Australian Capital Territory, neither of which being defined as states. It is thus not exactly accurate to refer to the current legal situation of Australia as a whole as permissive of assisted dying. However, for the sake of convenience and brevity, I will refer to this present case as the Australian example, taken to reflect the current situation in the states of Victoria, Western Australia, Tasmania, New South Wales, South Australia and Queensland, where VAD Acts have been implemented (Waller et al. 2023).

[17] See the 1997 Telecommunications Act from the Commonwealth Criminal Code https://www.legislation.gov.au/C2004A05145/2021-02-17/text Accessed June 10th 2024.

and founder of Exit International. He also attracted controversy because of his books which offer practical advice on end-of-life choices, be it through teaching methods on how to kill oneself peacefully or through advice on how to get to Switzerland to be granted suicide assistance.[18] Nitschke is perhaps better known in popular culture as the man behind the *Sarco*, a 3D-printed pod that can be flooded with nitrogen gas and ensure the death of its user within 10 min.[19] In effect, this means that physicians are strictly forbidden from discussing VAD during telemedicine consultations, over the phone or via e-mail. It is not difficult to see how such restrictions disproportionately affect rural requestors, as those are expected to be the main users of telemedicine. The communication ban, in addition to the right to institutional conscientious objection and the lack of available specialists in rural areas make it unreasonably difficult for those not living in metropolitan areas to access VAD (Close et al. 2021; Willmott et al. 2021; Furness et al. 2022; Sellars et al. 2022; Haining et al. 2023).

Let us now take a moment to summarize the key points of what we have just addressed here. In both Canada and Australia rural residents face disproportionate difficulties in accessing assisted dying because of the interaction between, on one hand, laws that protect conscientious objection (particularly of the institutional kind) and, in Australia, the legal barriers to discuss assisted dying as an end-of-life option via carriage services, and on the other hand the scarcity of specialists available outside of metropolitan areas and the low motility of the majority of aid-in-dying requestors. How does this situation fare with regards to our account of discrimination? What is evident is that those laws that disproportionately reduce access to rural residents were in no way designed to that end: it is the interaction between said laws and other factors that result in hindered access. This alone is enough rule out the possibility for direct discrimination: there is clearly no intent to prevent rural communities from accessing assistance. Perhaps what are dealing with here is a case of indirect discrimination. Recall our working definition of indirect discrimination: a policy or act is indirectly discriminatory against members of a group if the act reflects no bias on part of the discriminator against the group, the group is relevantly disadvantaged relative to other socially salient groups, the group allegedly discriminated against is a socially salient group and the suffered disadvantaged would not have occurred without past or present direct discrimination against the group in question. We have already established that the laws that create difficulties in access do not actually intent to do so, thus satisfying the first condition: laws that protect conscientious objection or forbid suicide talk over the phone do not reflect negative attitudes towards rural communities. Things however get a little muddier onwards. This is because, in order to be qualified as victims of indirect discrimination, rural communities must be understood as a socially salient group (and so must urban communities be as well if we are to satisfy the relevant relative disadvantage condition). Furthermore, it also must be demonstrated that rural

[18] https://www.exitinternational.net/about-exit/dr-philip-nitschke/ Accessed June 7th 2024.

[19] https://www.bbc.com/news/technology-59577162 Accessed June 7th 2024.

communities, if they qualify as socially salient, have been victims of previous directly discriminatory policies or acts. Let us start by examining how rural communities fare against the social salience condition.

Recall that according to Lippert-Rasmussen's account of discrimination and its constitutive features, a group is to be qualified as socially salient if perceived membership of it is important to the structure of social interactions across a wide range of social contexts (Lippert-Rasmussen 2014: 36). Note also that membership of the group does not need to be immediately visible for a group to be considered as socially salient: all that matters is that belonging to the group shapes and influences members' experiences and interactions in a wide range of contexts (ibid: 31). Following these premises, we are left with two questions: (1) before even considering social salience, can rural communities be said to belong to a unified group? And (2) does belonging to this alleged group actually structure the experiences and interactions of its members in a wide range of social contexts? With regards to the first question, it may reasonably be argued that individuals in rural communities are highly heterogenous and that just living in some place and not another is not enough to warrant the use of the group category. This is however a mistake, for similar arguments could be raised against the existence of other groups which are undoubtedly not only groups but socially salient ones. Think of women, blacks, people living with disabilities, for paradigmatic examples of groups that have been subjected to discrimination. All of these groups' members are widely heterogenous and yet no one would disagree with the fact that belonging to them (or one of them) structures the member's interactions in a wide range of social contexts. So, the heterogeneity argument proves lacking, at best. People living in rural communities constitute a group at least in the most basic sense of the word: they belong to the group of people living in rural areas because they share the same *P* property of living in rural areas. This has no normative weight and is akin to saying that people with blue eyes constitute a group based on the possession of the property of having blue eyes: we are still in the realm of analytical classification. We can now turn the social salience problem, which will start injecting some normative component into our classification. After all, while people with blue eyes may indeed constitute a group, they most certainly do not constitute a socially salient group in the sense understood by Lippert-Rasmussen, or us for that matter: having blue eyes does not structure one's interactions in a wide range of social contexts. And so, we can now ask the following question: is living in a rural area as irrelevant to social interactions and experiences as is having blue eyes? The answer appears trivial: of course not. The urban-rural gap is a well-known phenomenon and its consequences are wide and far-reaching across a variety of social dimensions. Take health for instance: beyond the lack of aid-in-dying availability we have addressed here, rural communities have worst cancer outcomes (Bhatia et al. 2022), have less opportunities for cardiac rehabilitation programs (Sangster et al. 2013) and are routinely neglected in epidemiological research (Smith et al. 2008). There are also significantly less economic and professional opportunities in rural areas, which translates into increased poverty and less efficient infrastructure than in urban settings (Young 2013). These are but a few of the ways in which rurality shapes the lives of rural communities and its members.

This is however not just a fact of social life. The spatial distribution of goods is itself a social product guided by social norms; a point that has been brought to the fore by theorists of spatial injustice as mentioned above (Pirie 1983). There is some level of prejudice against rural communities on the part of urban culture which has led not only to the bringing about of stigma but also to systematic neglect when designing and implementing policies of all kinds, health being a chief example (Mohatt et al. 2020). Another way to understand this sort of spatial inequality is to think not just in terms of prejudice against rural communities but of bias in favor of urban centers, which in turn has led to neglecting the needs and opportunities of those living far from them. This is not just mere accidental oversight: it is a structural feature of the social world, a structural injustice which was preventable but was not prevented due to a failure of equal concern for rural communities. As such, we start to see answers emerging regarding not only our inquiry about social salience but also about whether or not rural communities satisfy the fourth condition, that of having been or being victims of direct discrimination. It is now clear I believe that rurality shapes the social experiences of rural communities' members to such large extents that it can be safely argued that such a group meets the social salience criterion. I contend that the systematic neglect of such communities is enough to qualify the inequality and diminished opportunities they face as direct discrimination, particularly when considering that resources allocation has been biased towards urban centers, which as such logically form the relative socially salient group with regards to whom rural communities are disadvantaged.[20] But then, doesn't that contradict my assertion earlier in this section that hindered access to aid-in-dying for rural communities should be understood as indirect discrimination? Surely this hinderance of access is also due to neglect and bias towards urban centers, why then shouldn't it count as direct discrimination? This is because access to aid-in-dying has nothing to do with the allocation of resources and everything to do with how resources were *previously* allocated. Indirect discrimination arises because of the background of direct discrimination, understood as the systematic neglect in allocating resources. In this sense, with respect to access to suicide assistance, rural communities suffer because of a pre-existing flawed distribution of goods: this is the domain of direct discrimination. This of course exists within a larger context of structural injustice, a point to which we will come back to later on in the normative part of the book.

To summarize, I argue that rural communities constitute a socially salient group, are relevantly disadvantaged relative to another group (urban communities), have been victims of direct discrimination (through systemic neglect in the distribution of goods and the allocation of resources), and are victims of indirect discrimination in the case of hampered access to assisted suicide.

---

[20] A bit of nuance is warranted here. While it is true that the urban-rural gap is the product of unjust power relations and affects access to assisted suicide, it does not follow that urban settings are void of problems themselves. Cities can, within themselves, be the source and setting of spatial injustices and more precisely can also produce unjust inequalities of access to assisted suicide. For a seminal reference on spatial injustice in urban settings see (Harvey 1973) and for the specific question of access to assisted suicide in urban settings see (Fortin et al. 2024).

2.2.5 Low-Income Groups

The fact that requests for assisted suicide are not uniformly distributed amongst social classes is now well-established in the empirical literature on end-of-life options. At the same time, these empirical findings carried with them a tough blow for proponents of the slippery slope argument. Studies from Oregon have shown that the profile most likely to undergo assisted suicide was a white, highly educated male (baccalaureate or higher) with high socio-economic status, between the ages of 45 and 74 years old (Hedberg et al. 2003; Battin et al. 2007). A meta-analysis of 25 articles analyzing deaths from assisted suicide in the United States, Luxemburg and the Netherlands found similar results (Steck et al. 2013), and a study from Belgium shows similar patterns (Dierickxi et al. 2015). Data from Switzerland also found that high socio-economic status, high education, no religious affiliation and living alone were the highest predictors of death by assisted suicide, although there does seem to be some form of gender gap as women were more likely to die by suicide assistance than men, albeit not consistently throughout the years (Steck et al. 2014, 2016, 2018). Notwithstanding some small variations between jurisdictions, the picture painted by these results appears fairly consistent across all regions: assisted suicide is something that is requested and accessed by the most privileged and powerful strata of society and definitely not by the poor and vulnerable.[21,22]

These observations should leave us with quite a bit of questions, and it is important to clarify which ones are actually relevant to ask. One essential point to clear is that these empirical findings, as they stand now, are not enough to substantiate any normative claim about a potential pattern of injustice in access to assisted dying for low-income groups. What we have available so far are descriptive statistics that unfortunately say little about the potential causal mechanisms that create such skewed distributions. Just as we witness a strong relation between social class and effective use of assisted suicide, which can be accurately labelled as a form of inequality in access in the purely descriptive and thus non-normative sense, empirical studies have also shown for instance that in Switzerland Catholics are less likely to request suicide assistance than Protestants (Spoerri et al. 2010). As mentioned above, data also from Switzerland has shown that being female increases the likelihood of requesting assisted suicide (Steck et al. 2018).[23] We must be careful not to

[21] See also (Castelli Dransart et al. 2019) for a systematic review of reasons for requests and attitudes toward aid-in-dying.

[22] Recent reports from Canada have however highlighted that there may be occurrences of citizens requesting medical assistance in dying due to a lack of opportunity to access required medical care (Coelho et al. 2023). I explore the ethical underpinnings of this matter more deeply in Sect. 3.2.2.

[23] The gender gap in assisted suicide is also empirically demonstrated but remains unfortunately underexplored. Some have argued that this could prove morally problematic particularly with regards to assisted suicide for psychiatric reasons as these requests might stem from suicidal behavior (Nicolini et al. 2022). As of the time of writing, there is scant evidence available as to why the gender gap exists, and even less that has identified normatively problematic features of the gap. However, even if that were the case that some injustice could be raised, its relation with discrimina-

mistakenly infer injustice when all that can reasonably be identified are descriptive relations between one factor and an outcome. Naturally, inequality based on class or gender may strike us a potentially more morally problematic than inequality based on religious affiliation within Christianism. We however need more than just these correlations to actually kick off a proper normative investigation. With regards to class, a number of equally valid hypotheses could be raised as to why we witness unequal access. There is one aspect of the question that must be cleared out before proceeding any further, that is the actual cost of assisted suicide for requestors. The actual cost of assisted suicide for requestors is to be considered as negligible, with one notable exception for requestors living abroad and who wish to travel to Switzerland in order to die, where costs can range anywhere between 8200$ and 19,000$.[24,25] The low cost of assisted suicide is evidenced by the overall reduction in healthcare costs that legalization should engender, as has been posited and verified (Emanuel and Battin 1998; Trachtenberg and Manns 2017). With the exception of so-called "suicide tourism" to Switzerland, the cost of assisted suicide for requestors is not prohibitive; a point that is often raised in arguments against the legalization of assisted suicide as some have traditionally argued that its low cost relative to other end-of-life options such as palliative care will inevitably disproportionately draw low-income groups to choose suicide instead of prolonging life. As the data has shown, we can now safely assert that this is not the case. But then why do we observe this phenomenon? Perhaps low-income groups are less informed and lack the administrative know-how and social resources to effectively request suicide assistance; in other words, unequal access would be due to a lack of information on their part. Perhaps low-income groups tend to be more associated with immigrants who in turn tend to be more religious, a strong predictor of not requesting suicide assistance; this is some form of cultural argument. Or perhaps low-income groups tend to visit their general practitioner less often, thus learning of potential terminal conditions too late to have the time to request suicide assistance. These are all at best educated guesses and this is by no means an exhaustive list, we can however quickly point out that not all these hypothetical reasons are equally problematic normatively wise. In order to properly explore the unjust and potential discriminatory nature of this inequality, we need more than mere educated guesses and hypotheses.

As it turns out we do have more at our disposal, not much but enough to put us on a more substantial track. A recent study from Canada by Redelmeier and

tion as I frame it in this book would fall out of its scope, for I am here concerned with discrimination in *access* and not discrimination in *effect*. According to the framework I have adopted throughout this book, the relevant angle would thus be to explore if *men* are discriminated in accessing assisted suicide as opposed to women. As a matter of scope and relevance, I do not consider this question here.

[24] https://features.dignityindying.org.uk/true-cost-dignitas/ Accessed June 12th 2024

[25] On the topic of justice and the cost of assisted suicide, there is a small literature that provided arguments both in favor and against for-profit suicide assistance, or commercial assisted suicide. See (Kipke 2017; Cholbi 2023; Della Croce 2023).

colleagues took this linear distribution between social classes as a starting point and sought to provide an in-depth investigation of the phenomenon, through survey experiment and epidemiological data. What they found was not only that the effect of social class was again confirmed, but also that this effect was quantifiable and persistent even when adjusting for a variety of factors such as age, sex, location, and others: the odds of receiving suicide assistance decreased by 39% for patients in low socioeconomic groups when compared with patients in high socioeconomic groups (Redelmeier et al. 2021). This is especially concerning when one considers that support for assisted dying, or as we mentioned before MAiD in Canada, is as strong in Canadia poor households (below an annual income of 40,000 CAD) as it is in rich households (above an annual income of 100,000 CAD) (Reid 2016). This discrepancy is puzzling to say the least. Redelmeier and colleagues thus decided to perform a survey experiment to try and isolate the effect that social class had on physicians' assessment of patient suffering. What they found was that overall mean physicians' judgements of suffering were higher when assessing the pain of wealthy patients rather than that of poor patients. One key possible explanation for this unequal treatment highlighted by the authors is the so-called thick-skin bias. The thick-skin bias is a well-documented phenomenon in psychology that describes the tendency to overestimate the capacity to endure pain due to repeated hardships, thus leading to increased pain tolerance and reduced distress from negative life events. This specific cognitive bias has been shown to affect not only the poor but also racial minorities (Hoffman et al. 2016; Hoffman and Trawalter 2016; Cheek and Shafir 2024). We may thus conclude that reduced access to aid-in-dying from the poor may be due to some unconscious bias that leads physicians to take their suffering less seriously, especially since we know for a fact that the poor and the rich equally support assisted suicide.

Let us return one final time to our analytical account of discrimination and see how what have just seen fares against our definitions. As we are now familiar with, we must first establish that the group at hand is socially salient. In the case of the poor, this is easily settled: it is evident how low income structures the interactions and experiences of those belonging to low-income groups, be it through reduced access to basic necessities, failure of equal concern or lack of social respect towards the group. Just our little detour through the thick-skin bias is enough to show how belonging to the group of the poor structures one's interactions in society. We have also seen that in the case of access to assisted dying, the poor are disadvantaged relative to wealthier groups: their belonging to the group of the poor makes it more difficult for them to access assisted dying, all things considered. We must now figure out whether this is a case of direct or indirect discrimination. Here again, the answer is clear enough, for the thick-skin bias is by definition a bias, thus ruling it out of the indirect discrimination category (recall that for a policy or act to be indirectly discriminatory, it must reflect no bias on the part of the one that discriminates against the group at hand). Direct discrimination does not require its perpetrator to consciously hold a prejudice against the discriminated group. Unconscious biases may also be at the center of cases of direct discrimination: the P property that the discriminator believes (although unconsciously) that members of the group possess is that of high resilience due to their more difficult lives; this leads the discriminator to treat members of the group worse than the wealthier because they believe that the wealthier do not possess this P property; this P property is associated with the discriminated group; and access to assisted suicide is a relevant type of act in the sense that the thick-skin bias impacts the poor on a much broader variety of social situations. We thus have a textbook diagnosis of direct discrimination when looking at why low-income groups are less likely to access assisted dying, even when support for assisted death is uniformly distributed across all social classes. We must however remain cautious not to over-generalize our findings: more empirical research is required to properly elucidate the complexity of the issue, as the results I am basing this short conceptual analysis on are politically and culturally situated and may (but may also not) be context dependent. As we saw earlier and looking back at the uneven patterns of distribution of effective recourse to assisted dying and the relatively lower proportion of low-income individuals compared with their wealthier counterparts, one can for instance reasonably wonder if, perhaps, low-income groups lack the knowledge and practical information regarding how to access assisted dying. Or perhaps they, as we highlighted, are less confident in navigating the medical world than other richer citizens, be it in terms of money or social capital. What matters here is that this concrete uneven pattern of distribution can be morally problematic even in the absence of the thick-skin bias, although it looks like such other cases would fall under the heading of indirect rather than direct discrimination. What we have at hand is at least enough to kickstart philosophical investigation and situate the cases we are interested in within a conceptual mapping, although one can only hope for more empirical research inquiring into this very question to come up in the future.

2.3 Mapping Discrimination in Assisted Dying

We have now carefully assessed each one of our cases: disability, mental disorders, existential suffering, rural communities, and low-income groups. While the former three are usual suspects in discussing discriminatory access to assisted dying, the latter two are often overlooked, despite empirical evidence pointing to significant inequality of access. By applying Lippert-Rasmussen's analytical accounts of direct and indirect discrimination, we have now shed at least a little bit of light on what kind of discrimination, if any, is at stake in each of those cases. For the sake of clarity and ease of reading, let us summarize our findings as follows (Table 2.1):

Table 2.1 Discrimination type

Direct discrimination	Mental disorders	Low-income (Thick-skin bias)
Indirect discrimination	Disability	Rural communities
Nondiscrimination	Existential suffering	

Interestingly, both pro- and anti-assisted dying arguments based on disability fall into the category of indirect discrimination: neither the practical inability for some people with disabilities to effectively be assisted in suicide nor the disparate impact argument (or slippery-slope argument) can be traced back to intentional action on the part of legislators. People with mental disorders are, on the other hand, intentionally excluded from participating in assisted dying in most jurisdictions. Low-income groups have been documented to be suffering from bias on the part of healthcare professionals due to their supposedly greater resilience to pain, thus directly exercising a bias against them and being directly discriminated against, at least in this specific case. Rural communities have long suffered from indifference and structural neglect, which persists in being effectively able to access assisted dying. People with no condition other than existential suffering cannot be said to be discriminated against since they do not constitute a socially salient group in any form or shape. And as we have seen, children could also be included in the direct discrimination category. Reflecting Lippert-Rasmussen's distinction between direct and indirect discrimination, we can clearly see the deontic divide at play here between groups that are excluded from participation based on intention or due to oversight. Things are however slightly more complicated than just equating direct discrimination with intentionality and indirect discrimination with no intentionality. While this equation surely makes sense when looking at excluding access to mental disorders or failing to acknowledge the physical limitations of some people with disabilities, which amounts to them losing substantial access to assisted dying, things become muddy when looking at the case of low-income groups and the thick-skin bias. As we saw earlier, the thick-skin bias that leads practitioners to be less likely to offer the possibility for assisted dying to individuals from low-income backgrounds is, as the name suggests, a bias that ends up being clearly responsible for treating concerned individuals differently, arguably worse, based on some perceived property associated with being a member of that group: it is direct discrimination. But something seems slightly off in asserting that this latter case is conceptually equal to that of the exclusion of persons with mental disorders from participation.

We can isolate two reasons as to why that is the case: the first being normative and the second being conceptual. We will turn to the normative one over the course of the next chapter, but as we shall see it has to do with the potential for justification of said exclusion. Recall that the analytical account of discrimination we have used throughout this analysis is not moralized, meaning that it does not follow from concluding that an act is discriminatory that it is wrongful. On the normative dimension, these two cases differ insofar as while the exclusion of persons with mental disorders from assisted dying has some degree of potential for justification ("it is discriminatory, but it is for their best"), the same cannot be said for the pervasive exclusion of individuals from low-income backgrounds. Intentionality matters here, for it grants discrimination a rationale, a reason for which it may be beneficial to the discriminated group and as such be morally justifiable and even desirable. One can see it quite quickly: there is an element of deliberateness that is present in the exclusion of those with mental disorders that is absent in the case of those from rural communities. As such, one can call such distinction as of deontic nature. This deliberate component makes it possible for the exclusion to be, at least theoretically, positive, if it were for instance for the group's own good as we have mentioned just before. Of course, this does not imply that it is necessarily the case, and perhaps the fact of discrimination, while open to justification, may well be unjustified. The key normative difference here is that while such acts that intentionally discriminate may be morally *unjustified*, those that unintentionally discriminate are plainly *unjustifiable* in a very non-rhetorical way: it is logically impossible to provide any semblance of moral justification for these.

On the conceptual side of things, there is an obvious difference in the *locus* of discrimination that we have not addressed until now. While the exclusion of persons mental disorders occurs through legislation that sets up the boundaries of eligibility to assisted dying, which clearly and transparently lays out conditions for eligibility that exclude those with mental disorders, the exclusion of low-income groups seems to occur on a much more diffuse level: it occurs on individual interactions that are inevitably shaped by heuristics, stereotypes, and prejudice, all of which are significantly informed by the structural features of our social world (Koening and Eagly 2019). It is important here not to confuse the *locus* of discrimination with its *source*, for one could reasonably argue that, at the end of the day, legislation is not created out of nothing and most likely also reflects structural features of the social world. Perhaps the idea that persons with mental disorders, or at least a large enough portion of them, are not able to autonomously make decisions regarding their deaths is the result of deeply ingrained stereotypes about the nature of mental disorders and those who live with them. Or perhaps the oversights that have led people with severely limiting physical disabilities to be de facto excluded from assisted dying legislation find their source in systematic neglect and indifference towards the lives of those who are not able-bodied. These reasonable explanations are all structural in nature, and as such, one could make the reasonable claim that all sources of discrimination are ultimately structural. This is why the idea of a locus of discrimination is important, and we can here definitely distinguish between that of the legal and that of the purely structural. By purely structural, it must be understood that the stereotypes, prejudices, or indifference that have brought up unequal access *only* exist in the structural realm and have not been enshrined into legislation.[26] By drawing on this further distinction, and removing the case of existential suffering from our typology since, as we have previously seen, it does not actually qualify as discriminatory, we can reach the following classification (Table 2.2):

[26] Granted, the reader should be aware that while this book specifically addresses the cases of low-income groups and rural communities as illustrations of times where the locus of discrimination in accessing assisted dying, these are no more than select cases and as such are nowhere near exhaustive of all cases that share this similar property. The issue with such cases is that they are, by definition, harder to shed light on since they do not necessarily relate to any piece of legislation that seeks to regulate assisted dying. I have for instance previously argued that prisoners are also structurally disadvantaged in access to assisted dying for reasons ontologically similar to those mentioned here, see (Della Croce 2020, 2022).

Just like low-income groups, rural communities are not in any way legally excluded from the legally established criterion of eligibility for assisted dying. In the case of rural communities, the reason why access is unequal differs from that of low-income groups: it is not as much a display of stereotype as it is one of systematic indifference. From a strictly consequential point of view, the end result bears no difference for the locus of discrimination is the same: discrimination stems from structural factors that are in no way intentional inasmuch as they are not enshrined in law. Now where does that leave us normatively speaking? One will be quick to point out that this distinction between loci of discrimination reflects the distinction between justifiable and unjustifiable exclusion that we have seen before. Discrimination that happens through law is intentional and thus holds the potential for justifiability, whereas discrimination that fully happens through structures has no such potential. This however leaves the case of disability in a rather awkward position. This is because our previous analysis of intentionality naturally assumes that the kind of discrimination we are dealing with is of the direct kind. Direct discrimination, because of its component of intentionality, holds the possibility for being benevolent; indirect discrimination does not. As such, while the locus of discrimination that persons with disabilities face is indeed anchored in the realm of the legal, it remains that such exclusion can in no way be benevolent for it is unintentional in the first place. Thus, if one were to be more accurate in providing a mapping that includes the normative feature of potential for benevolence and justifiability, one would arrive at something along those lines (Table 2.3):

Doing so accounts for the unjustifiability of the exclusion of those with disabilities, despite the locus of discrimination being legal. What we will be interested in for the remainder of the book are nevertheless those cases where the locus of discrimination is purely structural, for such cases are usually left unaddressed in the ethics of assisted dying. What this mapping ultimately reveals is the pervasive and often underappreciated role of structural injustice in shaping access to assisted dying. Even where no explicit legal exclusion exists, structural dynamics—through prejudice, neglect, or geographic disadvantage—can produce discriminatory outcomes just as those produced by formal legal rules. If our goal is to secure just and equitable access to assisted dying, we must be attentive not only to what the law

Table 2.2 Locus of discrimination

	Legal	Structural
Direct discrimination	Mental disorders	Low-income (Thick-skin bias)
Indirect discrimination	Disability	Rural communities

Table 2.3 Justifiability of discrimination

	Justifiable	Unjustifiable
Direct discrimination	Mental disorders	Low-income (Thick-skin bias)
Indirect discrimination		Rural communities
		Disability

says, but also to how structural conditions shape its application on the ground. What we have also seen through this short mapping is that such cases of exclusion are never morally justifiable and are, by definition, much harder to detect than explicit, direct, cases of discriminatory access. As we will see in section 2, this does not mean that they should hold less importance in the ethics of assisted dying, quite the contrary. If our goal is to promote justice and secure just and equitable access to assisted dying, we must be attentive not only to what the law says, but also to how structural conditions shape its application. Our analysis along with this short mapping have further helped us in disentangling the different uses of discrimination in the ethics of assisted dying and hopefully made it a little clearer where the use of such terminology is merely rhetorical and where it is substantiated. Scholars and readers alike should thus exercise caution when engaging with the literature on assisted dying for it is unfortunately fraught with sensationalist terminology, as if one was hoping it would convey their point across better. It unfortunately does not, and not only does hoisting such terminology in cases where it is unsubstantiated philosophically erroneous, but it also has the effect of potentially conceptually obfuscating less visible cases of actual discrimination.

References

Altman, Andrew. 2020. Discrimination. In *The Stanford Encyclopedia of Philosophy* (Winter 2020 Edition), ed. Edward N. Zalta. https://plato.stanford.edu/archives/win2020/entries/discrimination/

Australian Institute of Health and Welfare. 2020. Rural and Remote Health. https://www.aihw.gov.au/reports/rural-remote-australians/rural-and-remote-health, accessed May 13th 2024.

Batavia, Andrew. 1997. Disability and Physician-Assisted Suicide. *New England Journal of Medicine* 336 (23): 1671–1673.

Battin, Margaret P., Ages van der Heide, Linda Ganzini, G. Gerrit van der Wal, and Bregje D. Onwuteaka-Philipsen. 2007. Legal Physician-Assisted Dying in Oregon and the Netherlands: Evidence Concerning the Impact on Patients in "Vulnerable" Groups. *Journal of Medical Ethics* 33:591–597.

Bhatia, Smita, Wendy Landier, Electra D. Paskett, et al. 2022. Rural-Urban Disparities in Cancer Outcomes: Opportunities for Future Research. *JNCI: Journal of the National Cancer Institute* 114 (7): 940–952.

Bickenbach, Jerome. 1998. Disability and Life-Ending Decisions. In *Physician Assisted Suicide: Expanding the Debate*, ed. Margaret P. Battin, Rosamond Rhodes, and Anita Silvers. Routledge.

Braun, Esther. 2023. An Autonomy-Based Approach to Assisted Suicide: A Way to Avoid the Expressivist Objection Against Assisted Dying Laws. *Journal of Medical Ethics* 49 (7): 497–501.

Braun, Esther, Matthé Scholten, and Jochen Vollman. 2024. Assisted Suicide and the Discrimination Argument: Can People with Mental Illness Fulfill Beneficence- and Autonomy-Based Eligibility Criteria? *Bioethics* 38 (1): 61 68.

Carter v. Canada (Attorney General). 2015. 1 S.C.R. 331.

Castelli Dransart, Dolores A., Sylvie Lapierre, Annette, et al. 2019. A Systematic Review of Older Adults' Request for or Attitude Toward Euthanasia or Assisted-Suicide. *Aging & Mental Health* 25 (3): 420–430.

Cheek, Nathan N., and Eldar Shafir. 2024. The Thick Skin Bias in Judgments About People in Poverty. *Behavioural Public Policy* 8 (2): 238–263.

Chhikara, Neelam. 2017. Extending the Practice of Physician-Assisted Suicide to Competent Minors. *Family Court Review* 55 (3): 430–443.

Cholbi, Michael. 2023. Envisioning Markets in Assisted Dying. In *New Directions in the Ethics of Assisted Suicide and Euthanasia*, ed. Cholbi Michael and Jukka Varelius, 2nd ed. Springer.

Close, Eliana, Katrine Del Villar, Lindy Willmott, and Ben P. White. 2021. Voluntary Assisted Dying and Telehealth: Commonwealth Carriage Service Laws are Putting Clinicians at Risk. *The Medical Journal of Australia* 215 (9): 406–409.

Close, Eliana, Ruthie Jeanneret, Jocelyn Downie, Lindy Willmott, and Ben P. White. 2023. A Qualitative Study of Experiences of Institutional Objections to Medical Assistance in Dying in Canada: Ongoing Challenges and Catalysts for Change. *BMC Medical Ethics* 24:71.

Coelho, Ramona, John Maher, and K. Sonu Gaind. 2023. The Realities of Medical Assistance in Dying in Canada. *Palliative & Supportive Care* 21 (5): 871–878.

Colburn, Ben. 2022. Disability-Based Arguments Against Assisted Dying Laws. *Bioethics* 36 (6): 680–686.

Coleman, Diane. 2010. Assisted Suicide Laws Create Discriminatory Double Standard for Who Gets Suicide Prevention and Who Gets Suicide Assistance: Not Dead Yet Responds to Autonomy, Inc. *Disability and Health Journal* 3 (1): 39–50.

Cossette-Lefebvre, Hugo. 2020. Direct and Indirect Discrimination. *Public Affairs Quarterly* 34 (4): 340–367.

Cowley, Christopher. 2016. A Defence of Conscientious Objection in Medicine: A Reply to Schuklenk and Savulescu. *Bioethics* 30 (5): 358–364.

De Keijzer, Sedona Celine, Guy Widdershoven, A. A. Eduard Verhagen, and H. Roeline Pasman. 2023. The Age Limit for Euthanasia Requests in the Netherlands: A Delphi Study Among Paediatrics Experts. *Journal of Medical Ethics* 49 (7): 458–464.

Del Villar, Katrine, Eliana Close, Rachel Hews, Lindy Willmott, and Ben P. White. 2021. Voluntary Assisted Dying and the Legality of Using a Telephone or Internet Service: The Impact of Commonwealth "Carriage Services" Offences. *Monash University Law Review* 47 (1): 125–173.

Della Croce, Yoann. 2020. The Great Escape? A Liberal Perspective on Assisted Suicide for Prisoners. *Ethics, Medicine and Public Health* 15:100566.

———. 2022. Assisted Suicide for Prisoners: An Ethical and Legal Analysis from the Swiss Context. *Bioethics* 36 (4): 381–387.

———. 2023. Against Commercial-Assisted Suicide. *Bioethics* 37 (7): 617–623.

Dierickx, Siegrid, Luc Deliens, Joachim Cohen, and Kenneth Chambaere. 2015. Comparison of the Expression and Granting of Requests for Euthanasia in Belgium in 2007 vs 2013. *JAMA Internal Medicine* 175 (10): 1703–1706.

Edwards, Steven D. 2003. Three Concepts of Suffering. *Medicine, Health Care and Philosophy* 6:59–66.

Eidelson, Benjamin. 2015. *Discrimination and Disrespect*. Oxford University Press.

Emanuel, Ezekiel J., and Margaret P. Battin. 1998. What are the Potential Cost Savings from Legalizing Physician-Assisted Suicide? *New England Journal of Medicine* 339 (3): 167–172.

Flamm, Michael, and Vincent Kaufmann. 2006. Operationalising the Concept of Motility: A Qualitative Study. *Mobilities* 1 (2): 167–189.

Fortin, Sylvie, Sabrina Lessard and Marie-Eve Samson. 2024. "Diversity and Access to Palliative Care and Medical Assistance in Dying in an Urban Setting." *OMEGA – Journal of Death and Dying* 0 (0): 1-20.

Furness, Kate, Jim Howe, Mitchell Chipman, Nirasha Parsotam, and Margaret O'Connor. 2022. The Commonwealth Criminal Code Restricts the Use of Carriage Services to Access Voluntary Assisted Dying in Victoria: A Perspective. *Australian Health Review* 47 (1): 64–66.

Gaignard, Marie-Estelle, and Samia Hurst. 2019. A Qualitative Study on Existential Suffering and Assisted Suicide in Switzerland. *BMC Medical Ethics* 20 (1): 34.

Gill, Carol J. 2010. No, We Don't Think Our Doctors Are Out to Get Us: Responding to the Straw Man Distortions of Disability Rights Arguments Against Assisted Suicide. *Disability and Health Journal* 3 (1): 31–38.

Golden, Marilyn, and Tyler Zoanni. 2010. Killing us Softly: The Dangers of Legalizing Assisted Suicide. *Disability and Health Journal 3* (1): 16–30.

Haining, Casey M., Lindy Willmott, and Ben P. White. 2023. Accessing Voluntary Assisted Dying in Regional Western Australia: Early Reflections from Key Stakeholders. *Rural and Remote Health* 23 (4): 8024.

Han, John. 2023. Terminalism and How Dying Patients are Conditioned as Docile Bodies. *Journal of Medical Ethics* 50 (2): 116–117.

Hartley, David. 2004. Rural Health Disparities, Population Health, and Rural Culture. *American Journal of Public Health* 94 (10): 1675–1678.

Harvey, David. 1973. *Social Justice and the City*. Edward Arnold.

Hatherley, Joshua J. 2019. Is the Exclusion of Psychiatric Patients from Access to Physician-Assisted Suicide Discriminatory? *Journal of Medical Ethics* 45:817–820.

Hedberg, Katrina, David Hopkins, and Melvin Kohn. 2003. Five Years of Legal Physician-Assisted Suicide in Oregon. *New England Journal of Medicine* 348 (10): 961–964.

Hinshaw, Stephen P., and Andrea Stier. 2008. Stigma as Related to Mental Disorders. *Annual Review of Clinical Psychology* 4:367–393.

Hoffman, Kelly M., and Sophia Trawalter. 2016. Assumptions about Life Hardship and Pain Perception. *Group Process & Intergroup Relations* 19 (4): 493–508.

Hoffman, Kelly M., Sophia Trawalter, Jordan R. Axt, and M. Norman Oliver. 2016. Racial Bias in Pain Assessment and Treatment Recommendations, and False Beliefs about Biological Differences Between Blacks and Whites. *PNAS* 113 (16): 4296–4301.

Kamisar, Yale. 1958. Some Non-religious Views Against Proposed Mercy-Killing Legislation. *Michigan Law Review* 42 (6): 969–1042.

Kipke, Rroland. 2017. Why not Commercial Assistance for Suicide? On the Question of Argumentative Coherence of Endorsing Assisted Suicide. *Bioethics* 29 (7): 516–522.

Kirby, Jeffrey. 2021. Should Institutional Conscientious Objection to Assisted Dying be Accommodated? *Canadian Journal of Bioethics* 4 (1): 15–20.

Knox, Michelle, and Adrian Wagg. 2023. Institutional Resistance to Medical Assistance in Dying in Canada: Arguments and Realities Emerging in the Public Domain. *Healthcare* 11:2305.

Koening, Anne M., and Alice H. Eagly. 2019. Typical Roles and Intergroup Relations Shapes Stereotypes: How Understanding Social Structure Clarifies the Origins of Stereotype Content. *Social Psychology Quarterly* 82 (2): 205–230.

Lippert-Rasmussen, Kasper. 2014. *Born Free and Equal? A Philosophical Inquiry into the Nature of Discrimination*. Oxford University Press.

MacKenzie, Jordan. 2023. Giving the Terminally Ill Access to Euthanasia is not Discriminatory: A Response to Reed. *Journal of Medical Ethics* 50 (2): 123–123.

Mohatt, Nathaniel V., Dennis Mohatt, Lorraine T. Benuto, Akihiko Masuda, William O'Donohue, and Melanie P. Duckworth. 2020. Rural Prejudice – Urban Bias: The Stories and Structures that Oppress Rural Communities. In *Prejudice, Stigma, Privilege, and Oppression: A Behavourial Health Handbook*, ed. Lorraine T. Benuto, Melanie P. Duckworth, Akihiko Masuda, and William O'Donohue. Springer.

Nelson, Ryan H., Brent Kious, Emily Largent, Bryanna Moore, and Jennifer-Blumenthal Baby. 2025. Is Suffering a Useless Concept? *The American Journal of Bioethics*. https://doi.org/10.1080/15265161.2024.2353799.

Nicolini, Marie E., Chris Gastmans, and Scott Y. H. Kim. 2022. Psychiatric Euthanasia, Suicide and the Role of Gender. *The British Journal of Psychiatry* 220 (1): 10–13.

Ogilvie, Alan D., and S. G. Potts. 1994. Assisted Suicide for Depression: The Slippery Slope in Action? *BMJ* 309:492.

Pirie, Gordon H. 1983. On Spatial Justice. *Environment and Planning A: Economy and Space* 15 (4): 465–473.

Pronk, Rosalie, Kirsten Evenblij, Dick L. Willems, and Suzanne van de Vathorst. 2019. Considerations by Dutch Psychiatrists Regarding Euthanasia and Physician-Assisted Suicide in Psychiatry: A Qualitative Study. *The Journal of Clinical Psychiatry* 80 (2): 19m12736.

Radden, Jennifer, and Jonathan Y. Tsou. 2024. Mental Disorder (Illness). *The Stanford Encyclopedia of Philosophy* (Fall 2024 Edition), ed. Zalta, Edward N. and Uri Nodelman. https://plato.stanford.edu/archives/fall2024/entries/mental-disorder/

Redelmeier, Donald A., Kelvin Ng, Deva Thiruchelvam, and Eldar Shafir. 2021. Association of Socioeconomic Status with Medical Assistance in Dying: A Case-Control Analysis. *BMJ Open* 11:e043547.

Reed, Philip. 2023. Discrimination Against the Dying. *Journal of Medical Ethics* 50 (2): 108–114.

Reid, I. 2016. *Eight in Ten Canadians Support Advance Consent to Physician-Assisted Dying*. Toronto: Ipsos Reid. https://www.ipsos.com/en-ca/news-polls/eight-ten-80-canadians-support-advance-consent-physician-assisted-dying.

Riddle, Christopher A. 2017. Assisted Dying & Disability. *Bioethics* 31 (6): 484–489.

———. 2023. Is Medical Aid in Dying Discriminatory? *Journal of Medical Ethics* 50 (2): 122–122.

Rodriguez v. British Columbia (Attorney General). 1993. 107 D.L.R. (4th) 342.

Rooney, William, Udo Schuklenk, and Suzanne van de Vathorst. 2018. Are Concerns About Irremediableness, Vulnerability, or Competence Sufficient to Justify Excluding All Psychiatric Patients from Medical Aid in Dying? *Health Care Analysis* 26:326–343.

Sangster, Janice, Susan Furber, Philayrath Phongsavan, Margaret Allmann-Farinell, Julie Redfern, and Adrian Bauman. 2013. Where You Live Matters: Challenges and Opportunities to Address the Urban-Rural Divide Through Innovative Secondary Cardiac Rehabilitation Programs. *Australian Journal of Rural Health* 21:170–177.

Sarbey, Ben. 2023. Assisted Dying Programmes are not Discriminatory Against the Dying. *Journal of Medical Ethics* 50 (2): 115–115.

Savulescu, Julian, and Udo Schuklenk. 2017. Doctors Have No Right to Refuse Medical Assistance in Dying, Abortion or Contraception. *Bioethics* 31 (3): 162–270.

Schiller, Catharine J. 2017. Medical Assistance in Dying in Canada: Focus on Rural Communities. *The Journal for Nurse Practitioners* 13 (9): 628–634.

Schuklenk, Udo, and Suzanne van der Vathrost. 2015. Treatment-Resistant Major Depressive Disorder and Assisted Dying. *Journal of Medical Ethics* 41:577–583.

Schuklenk, Udo, Johannes J. M. van Delden, Jocelyn Downie, Sheila A. M. McLean, Ross Upshur, and Daniel Weinstock. 2011. End-of-Life Decision-Making in Canada: The Report by the Royal Society of Canada Expert Panel on End-of-Life Decision-Making. *Bioethics* 25 (Suppl. 1): 1–73.

Sellars, Marcus, Ben P. White, Patsy Yates, and Lindy Willmott. 2022. Medical Practitioners' Views and Experiences of Being Involved in Assisted Dying in Victoria, Australia: A Qualitative Interview Study Among Participating Doctors. *Social Science & Medicine* 292:114568.

Shadd, Philip, and Joshua Shadd. 2019. Institutional Non-participation in Assisted Dying: Changing the Conversation. *Bioethics* 33 (1): 207–214.

Shavelson, Lonny, Thaddeus M. Pope, Margaret P. Battin, Alicia Ouellette, and Benzi Kluger. 2023. Neurologic Diseases and Medical Aid in Dying Laws Create an Underclass of Patients Based on Disability. *The American Journal of Bioethics* 23 (9): 5–15.

Silvers, Anita. 1998. Disability Discrimination and the Duty to Protect Otherwise Vulnerable Groups in *Physician Assisted Suicide: Expanding the Debate*, edited by Battin, Margaret P, Rosamond Rhodes and Anita Silvers. Routledge.

Sleeman, Katherine, and Iain Chalmers. 2019. Assisted Dying: Restricting Access to People with Fewer than Six Months to Live is Discriminatory. *BMJ* 367:l6093.

Smith, Karly B., John S. Humphreys, and Murray G. A. Wilson. 2008. Addressing the Health Disadvantage of Rural Populations: How Does Epidemiological Evidence Inform Rural Health Policies and Research? *Australian Journal of Rural Health* 16:56–66.

Sneddon, Andrew. 2006. Equality, Justice, and Paternalism: Recentreing Debate About Physician-Assisted Suicide. *Journal of Applied Philosophy* 23 (4): 387–404.

Spoerri, Adrian, Marcel Zwahlen, Matthias Bopp, Felix Gutzwiller, and Matthias Egger. 2010. Religion and Assisted and Non-Assisted Suicide in Switzerland: National Cohort Study. *International Journal of Epidemiology* 39 (6): 1486–1494.

Steck, Nicole, Matthias Egger, Maud Maessen, Thomas Reisch, and Marcel Zwahlen. 2013. Euthanasia and Assisted Suicide in Selected European Countries and US States. *Medical Care* 51 (10): 938–944.

Steck, Nicole, Christophe Junker, Maud Maessen, Thomas Reisch, Marcel Zwahlen, and Matthias Egger. 2014. Suicide Assisted by Right-to-Die Associations: A Population Based Cohort Study. *International Journal of Epidemiology* 43 (2): 614–622.

Steck, Nicole, Matthias Egger, and Marcel Zwahlen. 2016. Assisted and Unassisted Suicide in Men and Women: Longitudinal Study of the Swiss Population. *The British Journal of Psychiatry* 208 (5): 484–490.

Steck, Nicole, Christohpe Junker, and Marcel Zwahlen. 2018. Increase in Assisted Suicide in Switzerland: Did the Socioeconomic Predictors Change? Results from the Swiss National Cohort. *BMJ Open* 8:e020992.

Sumner, Leonard W. 2019. Institutional Refusal to Offer Assisted Dying: A Response to Shadd and Shadd. *Bioethics* 33 (8): 970–972.

Trachtenberg, Aaron J., and Braden Manns. 2017. Cost Analysis of Medical Assistance in Dying in Canada. *Canadian Medical Association Journal* 189 (3): E101–E105.

Tsou, Jonathan Y. 2013. Depression and Suicide are Natural Kinds: Implications for Physician-Assisted Suicide. *International Journal of Law and Psychiatry* 36 (5-6): 461–470.

Vandenhole, Walter. 2005. *Non-Discrimination and Equality in the View of the UN Human Rights Treaty Bodies*. Intersentia.

Varelius, Jukka. 2014a. Medical Expertise, Existential Suffering and Ending Life. *Journal of Medical Ethics* 40 (104): 107.

———. 2014b. On the Relevance of an Argument as Regards the Role of Existential Suffering in the End-of-Life Context. *Journal of Medical Ethics* 40 (2): 114–116.

———. 2015. On the Moral Acceptability of Physician-Assisted Dying for Non-Autonomous Psychiatric Patients. *Bioethics* 30 (4): 227–233.

Waller, Katherine, Katrine Del Villar, Lindy Willmott, and Ben P. White. 2023. Voluntary Assisted Dying in Australia: A Comparative and Critical Analysis of State Laws. *University of New South Wales Law Journal* 46 (4): 1421–1470.

White, Ben P., Ruthie Jeanneret, Eliana Close, and Lindy Willmott. 2023. The Impact on Patients of Objections by Institutions to Assisted Dying: A Qualitative Study of Family Caregivers' Perceptions. *BMC Medical Ethics* 24 (22): s12910-023-00902-3.

Wiebe, Ellen, Brian Sum, Michaela Kelly, and Mirna Hennawy. 2022. Forced and Chosen Transfers for Medical Assistance in Dying (MAiD) Before and During the Covid-19 Pandemic: A Mixed Methods Study. *Death Studies* 46 (9): 2266–2272.

Willmott, Lindy, Ben P. White, Marcus Sellars, and Patsy Yates. 2021. Participating Doctors' Perspectives on the Regulation of Voluntary Assisted Dying in Victoria: A Qualitative Study. *The Medical Journal of Australia* 215 (3): 125–129.

Wright, Megan S. 2021. Equality of Autonomy? Physician Aid in Dying and Supported Decision-Making. *Arizona Law Review* 63:157–197.

———. 2023. Current Medical Aid-in-Dying Laws Discriminate Against Individuals with Disabilities. *The American Journal of Bioethics* 23 (9): 33–35.

Young, Alwyn. 2013. Inequality, the Urban-Rural Gap, and Migration. *The Quarterly Journal of Economics* 128 (4): 1727–1785.

Chapter 3
Structural Injustice in Assisted Dying

Abstract This second chapter builds upon the conclusions from Chap. 2 and begins by highlighting the conceptual relation between discrimination and structural injustice as well as vulnerability. It then fleshes out the essential yet often implicit role that structural injustice plays in many arguments in the ethics of assisted dying, especially so in those that oppose legalization. Three cases where structural injustice interfaces with assisted dying are then presented. The first two are those of life devaluation and expressive harm, and that of inequality of access to palliative care and choosing when options are limited. Following the same logic as that of Chap. 2, I then present a third kind of occurrence where structural injustice acts as a barrier to access to assisted dying, departing from the views of most common arguments which usually understand structural injustice as a force to acknowledge which provides reasons to oppose assisted dying. I contend that in many cases, some empirically demonstrated and some more speculative, structural injustices can often limit one's legitimate access to assisted dying.

Keywords Structural injustice · Vulnerability · Assisted dying · Palliative care · Disability · Equality

3.1 From Discrimination to Structural Injustice

We have now talked quite a bit about discrimination and highlighted a few cases where discrimination occurs in assisted dying and sketched out ways in which that they could potentially be morally problematic. We drew special attention to cases where discrimination did not occur via explicitly exclusionary legislation but rather via indifference or individual interactions, both of which are shaped and brought about by social structures. We also saw that, unlike cases of discrimination through legislation, discrimination that occurs in such diffuse ways is hard to justify: while there might be justifiable reasons for discriminating against persons with mental disorders in accessing assisted dying, one can hardly think of a morally sound justification for discriminating against low-income groups because of physicians'

Y. Della Croce, *Assisted Dying, Discrimination, and Structural Injustice*, SpringerBriefs in Ethics, https://doi.org/10.1007/978-3-032-11093-0_3

implicit bias. The intentionality of legislation makes it so there is room for justifying discrimination, which is not the case when discrimination is purely structurally induced. Intuitively, we may get the idea that these latter cases somehow relate to the idea of structural injustice. But how does structural injustice unfold in assisted dying, and what role should it play for the ethics of assisted dying? This third chapter aims to provide answers to both of these questions.

Discrimination and structural injustice are not conceptually distant, nor are they completely unrelated. It is here useful to start from the beginning and briefly shed light on the conceptual relation between discrimination and *oppression*. According to some critics of discrimination as a concept, most notably Iris Young, inquiring about discrimination in itself is a somewhat pointless endeavor, for the correct concept to address group-related injustice is oppression, to which discrimination is but just a derivative (Young 1990, 195). Such a view contends that by focusing too much on discrimination, a mostly agent-oriented concept, we run the risk of placing too much importance on interindividual relations and interactions at the cost of structural and institutional frameworks of oppression, which in the end are what we should really be addressing in order to address injustice (ibid). In his account of discrimination that we have made use of throughout this book, Lippert-Rasmussen addresses these critics and contends that his use of discrimination and Young's use of oppression reflect little more than terminological variation, for they essentially seek to capture the same phenomena (Lippert-Rasmussen 2014, 49). I endorse such a view and agree with Lippert-Rasmussen that, while similar, there is value in proposing a fully-fledged analytical account of discrimination that should be understood to complement the concept of oppression and not override it. While fascinating, this book is however not the adequate place to discuss the philosophical technicalities that shape the disagreements between theorists of discrimination and those of oppression. What matters to us is that these two concepts are largely thought to be related, and it does make intuitive sense that they should be. But let us pause quickly and highlight what exactly is meant by the idea of oppression and how it relates to the idea of structural injustice. In Young's terms, oppression is to be understood as an institutional constraint on self-development and expresses itself through a variety of what she refers to as the faces of oppression: marginalization, exploitation, cultural imperialism, violence, and powerlessness (Young 1990). Oppression then, like domination, is actually the concrete display of structural injustice: that is the actual realization of structural injustice. Structural injustice is defined as follows:

> Structural injustice, then, exists when social processes put large groups of persons under systematic threat of domination or deprivation of the means to develop and exercise their capacities, at the same time that these processes enable others to dominate or to have a wide range of opportunities for developing and exercising capacities available to them. Structural injustice is a kind of moral wrong distinct from the wrongful action of an individual agent or the repressive policies of a state. Structural injustice occurs as a consequence of many individuals and institutions acting to pursue their particular goals and interests, for the most within the limits of accepted rules and norms. (Young 2011, 52).

As Maeve McKeown has accurately highlighted, the idea that some injustices are embedded in structures is in itself not new and may be traced back at least to Marxist theory (McKeown 2021). The threats mentioned by Young find their origins in a variety of social processes: stigma, sexism, racism, ableism, capitalism, and so on. These processes are then incorporated into institutions and policies, at which point they become properly structural and not merely social. Since the source of structural injustice is to be found not in the easily identifiable actions of specific individuals but rather in diffuse social and institutional processes, ascribing responsibility for structural injustice is a difficult endeavor, a point also highlighted by Young (1990; Browne 2024).[1] Structural injustice finds its origins in social structures, rather than the isolated actions of individuals, which Young defines as the accumulated outcomes of the actions of the masses of individuals, enacting their own projects, often uncoordinated with many others (Young 2011, 62). As Jonathan Wolff put it, following this account of social structures, structural injustice is thus a set of collective unintended consequences which affects the opportunities available to different individuals (Wolff 2024, 14).

To better illustrate our point and highlight the relevance of thinking in terms of structural injustice in the context of assisted dying, it is useful to take a brief detour through the idea of vulnerability. The literature on vulnerability is, like pretty much all of the concepts we have made use of so far, rich both in content and disagreement. For the sake of brevity, we shall skip most of these disagreements and content ourselves with a synthetical account brought forward by Angela Martin and colleagues: vulnerability is to be defined as one's increased risk of being unjustly wronged or being unable to claim what one is legitimately entitled to (Martin et al. 2014).[2] Recall the alleged discriminatory cases we analyzed earlier. When adding the analytical layer of vulnerability to what we have seen in Chap. 2, it appears clear that there is some significant overlap between cases that qualify as discrimination and the alleged vulnerability of the group that is discriminated against.[3] The mentally ill, the poor, those living outside of cities and away from medical centers, or the disabled all intuitively seem to be, due to their membership in their own respective group, at higher risks of being morally wronged or being denied / unable to claim the opportunities they are entitled to. This is rather straightforward to see when considering assisted dying. The idea of the slippery slope is well attuned to the idea of some groups being at a higher risk of being wronged, and the overall issue of access that has constituted the common thread of this book so far is directly concerned with being unable to claim opportunities that one should legitimately be

[1] A complete discussion on the proper ascription, or the possibility of ascription, of responsibility in the case of structural injustice is beyond the scope of the present investigation, see (Aragon and Jaggar 2018), (Atenasio 2019), (Hayward 2017) and (Sangiovanni 2018) for extensive discussions of the matter.

[2] Normatively, this conceptual diagnosis correlates with an ethical duty to especially protect the interests of the vulnerable. See Tavaglione et al. (2016) for an overview of such duty.

[3] While this conceptual connection seems to point in the direction of an overlap between the category of vulnerable groups and socially salient groups, note that the former is a subset of the latter: all vulnerable groups are socially salient but not all socially salient groups are vulnerable. More nonetheless needs to be investigated regarding the relationship between the two.

entitled to, especially when access is impaired due to purely structural factors. In contrast, we are quick to see that the case of existential suffering that we concluded to be nondiscriminatory also seems at odds with vulnerability, for it is at best unclear how existential suffering makes one more likely to be wronged or have their legitimate opportunities constrained.

This is nice, but how exactly does that relate to structural injustice? The first thing to note about vulnerability is that it is not a property that is inherent to things: all of us have the potential to be vulnerable. Vulnerability is something that is ascribed to some more than others and as such, is both relational and something that is circulated. According to Jade Schiff, vulnerability is a problem of structural injustice because of how it is circulated, that is in a way that systematically benefits those in power and hurts those that are less privileged (Schiff 2024, 142). Just like we have seen with discrimination, structural injustice helps us lift off the idea of vulnerability from the individual to the social and understand that beyond the actions of individuals there are structures that foster discrimination and increase vulnerability and thus decrease one's ability to exercise rights they legitimately hold. If we look back at our rural communities and low-income groups which are structurally disadvantaged when it comes to access to assisted dying, it appears clear how these ideas are actually relevant to the ethics of assisted dying. Nevertheless, little attention has been given to such cases, let alone to structural injustice in assisted dying. In what follows, I address two the ways in which structural injustice concretely interfaces with assisted dying and then propose a third way that is almost completely absent from the literature, which focuses precisely on the case of structural injustice limiting access to assisted dying.

3.2 Three Occurrences of Structural Injustice in Assisted Dying

Much like the idea of discrimination, the idea of structural injustice permeates the discussions about the ethics of assisted dying. However, unlike discrimination, the usage of the terminology and language of structural injustice is almost always implicit and rarely openly laid out. Traditionally, the debate between pro- and anti-legalization has often been framed respectively as a liberal (at times libertarian) vs social justice strife. The former, that is the pro camp, usually rests their case on the grounds of individual choice and liberty whereas the latter takes on a more socially conscious approach that situations the individual within a web of power relations that make the idea of a fully autonomous choice fictitious at best.[4] It is thus only

[4] Interestingly enough, very few liberal political philosophers have actually sought to propose a fully-fledged philosophical stance in favor of assisted dying, despite prominent liberal (and in the case of Nozick, libertarian) philosophers such as John Rawls, Thomas Scanlon, Ronald Dworkin, Thomas Nagel and Judith Jarvis Thomson communicating, in 1997, an openly political stance in favor of assisted dying to the US Supreme Court in the context of the *Vacco v. Quill* and *Washington*

natural to find most of the familiar ideas constitutive of structural injustice theory in the arguments of those that oppose legalization or those who staunchly defend limited access and fight against expanding the boundaries of assisted dying. I here want to distinguish between two different ways in which the concept of structural injustice is implicit in two separate anti-assisted dying arguments. While I contend that the idea of structural injustice indeed permeates most of the argument against legalization or legislation expansion, this does not imply that these arguments are conceptually equal nor that they rely on the same kinds of structural features and injustices to bring their point across. These arguments are respectively that of the alleged devaluation of life and expressive harm that is suffered by some of the most vulnerable members of society, chiefly people living with disabilities, and that of the absence of a truly autonomous choice due to a lack of adequate medical alternatives to assisted dying, chiefly through the lack of available palliative care.[5] While the former is somewhat relational in nature and the latter is more concerned with distributive justice, both of these arguments somehow rely on structural injustice to gain normative traction. Following the same method we used through Chap. 2 when discussing discrimination, I then want to show that by addressing the ethics of assisted dying *from* the domain of the theoretical and only *then* go to the practical, in direct opposition to what is usually done in the literature, we are able to uncover some less obvious cases where structural injustice and assisted dying interface. In this very case, what I argue is that structural injustice is also relevant for those in favor of legalization or legislation expansion, for there is an untold story here that largely echoes what we saw when we conceptually mapped discrimination in assisted dying: that there are some groups that are structurally disadvantaged as far as their access to assisted dying is concerned, and these disadvantages are displays of pre-existing structural injustices. We'll move on to this final point once we've evidenced the conceptual relation between our two aforementioned anti-arguments and structural injustice and how they substantially rely on this very idea.

v. Glucksberg cases via a co-signed brief (Dworkin et al. 1997). Dworkin is the only key figure in contemporary liberal theory to have substantially argued in favor of assisted dying on the basis of liberty and equality, see (Dworkin 1994, 2000).

[5] Note that people living with disabilities do not constitute the entirety of what these arguments usually refer to as "the most vulnerable", just like the lack of available palliative care does not exhaust the lack of alternatives that individuals should allegedly have at disposition in order to make a truly autonomous choice. With regard to the former, the elderly are also a part of the population that is often considered greatly vulnerable when it comes to having their lives devalued, and one could imagine that lack of affordable healthcare could also reduce one's alternatives and nudge them into demanding assisted dying. I use these cases because they are those that are by far most often brought up in the literature, but one should keep in mind that they are by no means exhaustive.

3.2.1 Devaluation of Life and Expressivism

Let us first start with the most well-known argument anti-legalization and flesh out the role that structural injustice plays in giving substance to it: that of the *devaluation of life* argument. We have already touched on this idea earlier, and it serves as the basis for slippery slope arguments against the legalization of assisted dying.[6] To restate it, those who champion such a view claim that the lives of the most vulnerable, chiefly people with disabilities, are considered less valuable, or even burdensome, compared to those of other able-bodied citizens. As such, the risks of legalizing assisted dying are multiple. People with disabilities may have internalized the idea that their lives are somehow less worth living than those of others leading them to think they're a burden to others and society and ultimately leading them to choose assisted dying while they did not have the autonomy to make such a choice truly autonomously. Or perhaps the devaluation of their lives is too deeply ingrained in physicians" minds, which may lead them to propose and pressure people with disabilities to choose assisted dying, believing it is just better for them and society at large as their lives are allegedly miserable and costly. Looking back at our short overview of structural injustice, we can clearly see how the dynamics of pervasive ableism may end up oppressing those with disabilities and coercing them into hastened death, at least according to this variety of argument.[7] This is a textbook case of what we are trying to tackle here: there is no denying that those with disabilities live under constant threat of domination or oppression, defined as a restrained capacity for self-development, and that the origins of such threats are to be found in ableism, a deeply structural feature of Western societies. Here we see the aforementioned, and as I contend slightly exaggerated, divide between liberal and social justice perspectives on the ethics of assisted dying. While the liberal (arguably libertarian), championing the ideal of negative liberty, would likely defend the idea that the choice to request assisted dying is of strict concern only to the individual and little more, the more social justice-oriented theorist would answer that unfortunately things in the real-world are not so simple and that individuals live in webs of power relations and structural injustices that may thwart their judgments or those of others. This divide has largely shaped the ethics of assisted dying for decades.

Another, related but conceptually distinct, way in which this argument has been put forward is through what is referred to as *expressive* harm. The idea of expressive harm is present in a significant number of relational approaches to equality and justice and essentially states that aside from the practical, potentially harmful,

---

[6]As this chapter aims to clarify the relation between structural injustice, I do not proceed in any evaluation of the quality or relevance of the arguments presented. Nevertheless, see Battin et al. (2007) and Emanuel et al. (2016) for empirical evidence demonstrating the absence of a slippery slope after legalizing assisted dying.

[7]I have in the past proposed an argument that accepts the same premises regarding structural injustice and oppression but argues against the conclusions presented here; see Della Croce (2020).

consequences of a given policy, the *content* of the policy may also harm social groups affected by the policy strictly in virtue of the message it conveys to them. Think for instance of a discriminatory law that forbids people of a certain race to access public pools, because they are perceived to be unhygienic due to stereotypical assumptions that stem from deeply ingrained racism. Assume also that belonging to this certain race qualifies as membership of a socially salient group, for the sake of consistency with our previous account of discrimination. Should you be a member of this race, there is naturally a wrong insofar as your opportunities are concerned: your opportunities have effectively been unjustly reduced compared to those of others, and this is wrong. Calling back to the caricatural liberal egalitarian we've discussed just above, this would most likely be the end for them: in order to fix the wrong committed here, the unjust policy must be lifted and all will be well. But there is more at play here. According to relational egalitarians, while your opportunities certainly have been reduced and this of course matters, there is a pernicious other kind of a moral wrong associated with this discriminatory policy. It can indeed be contended that aside from the very pragmatic consequence of you not being able to access public pools, there is something deeply degrading and disrespectful with targeting your group of people and explicitly casting them aside with regard to something other races are still allowed to do. The discriminatory policy not only reduces your opportunities, but it also *expresses* contempt and disrespect to you: it makes you unequal in terms of opportunities and signals that you are in fact not of the same moral worth as your fellow co-citizens, at least according to the State (Schemmel 2012; Voigt 2018). It makes sense why this line of argument is usually nowadays associated with relational accounts of egalitarianism and justice for, unlike traditional purely distributive accounts of equality and justice championed by some liberal theorists, its normative focus lies in how citizens relate to each other in terms of equal standing, making such accounts much more attuned to structural injustice and dynamics of oppression and domination (Anderson 1999).[8] The expressivist argument is not in itself new and has arguably been a prominent component of disability studies. Traces of the expressivist argument go well beyond assisted dying and how it allegedly sends a demeaning message towards those with disabilities by signaling their lives as of less worth than those of others and can be found all the way back in debates about the use biotechnologies to prevent the birth of disabled infants and how such prevention fuels eugenics (Hofmann 2016; Gyngell and Douglas 2018). In both devaluation of life and expressivist arguments, the alleged harm done to the disabled finds its origins in pervasive ableism and its negative consequences to those concerned, which fully pertain to the realm of structural injustice. As such, despite often being implicit, both of these arguments are actually based in structural injustice and how assisted dying either fosters harm or offense to

[8] There would be much more to say regarding how relational conceptions of equality can contribute to the ethics of assisted dying as opposed to distributive ones, be it in terms of conceptual clarification and normative ideal. This is unfortunately not the place to do so, but one can hope that more work concerning equality and its demands in assisted dying sees the light of day in future developments in the literature.

a class of people which is already socially and politically disadvantaged. It follows from this line of thought that if we were to succeed in getting rid of ableism and guarantee those with disabilities proper equal standing in relation to both the State and fellow citizens, then assisted dying would be permissible. But as long as this world is not our world, we should refrain from legalization.

3.2.2 Access to Palliative Care and Choosing When Choices Are Limited

There is a second way in which structural injustice interfaces with assisted dying that is less concerned with pre-existing social relations and logics of disrespect and more with the distribution of opportunity that shape one's ability to actually make a choice deemed to be autonomous. The idea, broadly, is that lack of access to adequate medical care, in particular palliative care, may lead individuals to "choose" assisted dying because they did not have actual access to another option they might have preferred. This is thought to typically impact medical deserts the worst, but also any community that has systematically been neglected in healthcare availability, and usually such deprivation arises from structural injustices and segregation that range from classism to racism. This is distinct from the idea that some individuals may choose not to request assisted dying if treatment for the condition actually existed. What we are dealing with here are cases where other end-of-life options, most relevantly palliative care, are not available in a given community. To illustrate, imagine Bernard, a 60-year-old steel factory worker who is suffering from late-stage lung cancer with poor 6 months prognosis. Bernard is by no mean a high earner and lives in a working-class suburb of his city. He is now unable to work and knows that he will pass away before the end of year, and as such he has started to look into end-of-life options to make his last months as comfortable as possible, in agreement with his general practitioner and oncologist. Unfortunately, the availability of palliative care is scarce and far from his community, and to make things worse his medical insurance will not fully cover the costs incurred by his palliative care. Bernard being a husband and father of three children, to whom he is very attached and wishes to remain close by them, making the palliative care option far into the city unattractive. Furthermore, he does not want to leave debts to his family after his passing, since he knows his insurance will not cover all his costs and he may rack up medical bills that will be made part of his estate for his loved ones to possibility inherit. Knowing that without adequate care his last months will be extremely painful, he requests assisted dying, for which he is quickly approved and dies shortly after. While Bernard was certain that he did not want to endure the last months of his illness without care, he would have preferred this option as spending more time with his family was highly valuable to him and, as political liberals would say, to his conception of the good life. The challenge that is raised here is the following: was Bernard's choice to demand assisted dying truly autonomous[9]?

The aim of the upcoming lines is not to provide an answer to this challenge but rather to make explicit how such an argument, just like the aforementioned devaluation of life and expressivist arguments, actually relies on the idea of structural injustice. Here again, similar to the previous argument we reviewed, those who endorse such a view are actually not completely opposed to assisted dying. In the devaluation of life case, its supporters claim that only once we get rid of oppressive social relations that lead people to believe the lives of others (or their own in cases where such oppression is internalized) are less valuable and worth living if they do not fit some model of able-bodiedness (or youth in the case of the elderly), then and only then can we think about authorizing assisted dying. Here we are met with a similar, although arguably less demanding, argumentative logic: only once wide access to palliative care has been guaranteed can we legalize assisted dying. Proceeding in this way is, according to supporters of this view, the only way to actually guarantee autonomous choices at the end-of-life.[10] The relationship between assisted dying and palliative care, both philosophically as well as in concrete clinical practice, has been complicated and at times tumultuous (Gerson et al. 2020), and has arguably deeply reshaped the practice of palliative care where assisted dying has been legalized (Mathews et al. 2021). Some have even argued that the legalization of assisted dying actually runs the risk of impeding the development of palliative care, in this specific case in Canada (Collins and Leier 2017), a claim that has been partly rebutted in Europe (Chambaere and Bernheim 2015). What matters here is thus that of choice under suboptimal conditions, or choices where relevant alternatives are limited. What makes such arguments structural is the fact that the conditions under which choices are made are not randomly generated and, as mentioned above, are displays of logics of neglect regarding the provision of healthcare. It is already unfortunately well-established that social inequality impairs access to palliative care, be it through the lack of geographical availability for underserved community or due to financial constraints (Reimer-Kirkham et al. 2016; French et al. 2023). To reuse the liberal terminology we briefly touched on earlier, autonomy is here understood not to simply have one's choice unconstrained by external barriers (as traditional negative accounts of liberty would champion), but rather for one to be able to make a choice that is *meaningful*, that is to actually have a access to a range of options.[11] In this here case, the range of options according to which one's choice is made is the product of socio-political and economic relations of power that

[9] For an in-depth analysis of the ethics of such choices made in suboptimal conditions, see Wiebe and Mullin (2023) and Kious (2024).

[10] For extensive reviews, and rebuttals, of such views, see Barutta and Vollmann (2015) and Colburn (2025).

[11] This argument is somewhat transposable to the case of disability we saw just earlier in 2.2.1. The idea is that, aside from the devaluation of life and expressivist arguments that we touched on, disdain towards those with disabilities leads the State to provide assisted dying in lieu of other alternatives that those with disabilities would welcome. Assisted dying is here seen as an insincere way to

systematically leave some at a disadvantage, hence the essential role of structural injustice in such arguments.

3.2.3 A Forgotten Face: Structural Barriers to Access

We have now covered the devaluation of life, expressivist, and limited choices arguments and highlighted how each of those, in one way or another, rely on the idea of structural injustice to gain normative traction. In addition to their reliance on structural injustice, these arguments also share their anti-assisted dying normative positioning. Granted, this again reflects the (again, slightly caricatural) divide we've highlighted before, that between the liberal who champions individual liberty above all and the social justice-oriented theorist who is more likely to take the complexity of social phenomena into account before handing out their normative judgment. The former usually being in favor of assisted dying, and the latter against. It makes intuitive sense why the champion of individual liberty would fight tooth and nail in favor of assisted dying. It makes less intuitive sense why being attentive to social and structural dynamics, particularly structural injustice, should naturally lead to the conclusion of being against assisted dying. This is because it does not.

I now want to introduce an often-forgotten face of structural injustice in assisted dying, that of structurally imposed barriers that impair legitimate and rightful access to it. Recall that our working account of vulnerability highlights one's heightened risk to be unjustly wronged but also to be less able to claim what one is legitimately entitled to (Martin et al. 2014). It is here to the latter part of this definition that we should direct our attention: what of cases where someone has a right to access assisted dying but fails to do so due to structural constraints? If assisted dying is a right, be it grounded in self-determination, autonomy, or dignity, then failure for some to access it due to structural injustice is a failure of justice in the same sense as we have seen above. Why this very kind of injustice has so far received little attention in the ethics of assisted dying has to do with the meta framework in which the debate has taken place. Recall that, early in the book, we distinguished between dichotomic and continuum approaches in the ethics of assisted dying. To freshen our memory, arguments in the dichotomic approach are concerned with whether assisted dying, in any form, should be legalized or not. Arguments in the continuum approach instead assume that assisted dying is already legal and seek to tackle the question of its boundaries, i.e. who should be eligible and for what reasons. It is unsurprising that most of the arguments in the literature belong to the dichotomic approach, chiefly because of how long ethical quarrels about assisted dying have been going on in comparison to the recent growing of jurisdictions which have legalized it, prompting more exchanges about its ethical boundaries as opposed to

address the claims of the disabled and fails to provide with them with any meaningful options aside from voluntary death.

moral justification. This has resulted in the tension that I wish to shed light on in this chapter, making it seemingly normatively incompatible to simultaneously combat structural injustice and be in favor of assisted dying. This needs not be the case, especially not as far as the continuum approach is concerned.

In this final case, the way in which structural injustice interfaces with assisted dying is through what we will refer to as structural barriers. Structural barriers are those barriers that limit access due to patterns of disadvantage built into the ordinary functioning of institutions, policies, or social practices. An incompetent physician that, for some reason, is completely unaware that assisted dying is even an available option as such does not constitute a structural barrier. Structural barriers are the product of structural injustices that systematically disadvantage some social groups and increase their overall vulnerability, in this present case as far as their capacity to claim their legitimate rights is concerned. Think of formal versus substantial ideas of equality to illustrate what is at stake here: structural barriers create cases where, despite being formally equal, some are in practice less able to claim what they should be owed, making them substantially unequal. Recall the two cases of discrimination due to structural factors we have seen in Chap. 2: those of rural communities and low-income groups. What these two groups face are in fact structural barriers that hinder their capacity to claim their right to assisted dying, in the form of spatial injustice for the former and bias for the latter.

Most importantly, there is a crucial difference in how works that seek to investigate structural barriers to access assisted dying proceed in comparison to the above-mentioned structurally based arguments: those start from the empirical data and documented practices in how assisted dying is accessed rather than from purely theoretical and speculative standpoints about how it may be accessed. Some words of caution are required here in order not make overreaching claims. By starting from the empirical data, one should be clear that unfortunately there is very little data available that investigates causality between structural injustice and access to assisted dying. The data that is referred to here essentially consists of descriptive statistics, where being a victim of structural injustice is at best a predictor of reduced access to assisted dying. This does not rule out that in some cases, structural injustice acts as a proxy for something else which is the actual cause for unequal access. Such data should nevertheless raise some red flags and, at minimum, encourage further investigation into the precise causes of unequal access. As an illustration of contemporary research that has gone in that direction with troubling results, it has been shown that, at least in Québec, people with lower socioeconomic status, despite having poorer knowledge of end-of-life practices, saw assisted dying in a positive light, thus bolstering the idea that it is structural barriers that provide explanation for their underrepresentation and not disinterest or negative judgment (Bérubé et al. 2025). In Chap. 2, we saw that assisted dying is in fact something that is significantly more requested and accessed by the rich and highly educated, a thread common in all jurisdictions where it is permitted (Ganzini 2014; Steck et al. 2013, 2014, 2016, 2018).[12]

It is here useful to preemptively engage with a line of argument that those against assisted dying may come up with, which would consist in turning the argument the other way around: maybe there is no such thing as structural barriers but structural injustices and logics of oppression like ableism, racism, sexism, or classism, may increase one's risk to non-autonomously choose assisted dying. We can safely proactively reject such arguments for the data has simply proven the opposite, there is no such thing as an increased prevalence of assisted dying amongst the disabled, the poor or the racialized.[13] As we mentioned previously, while descriptive data is not necessarily enough to make precise inferences as to what is the source of unequal access (only that there is unequal access and that it has to do with factors usually associated with structural injustice), it is however enough to rule out any claim that structural injustices increase one's tendency to be assisted in dying. Unfortunately, such erroneous claims are often reported in the media in order to delegitimize assisted dying with very little philosophical or empirical basis. Perhaps most problematic is that such false accusations that assisted dying nudges the structurally vulnerable to request it based on socioeconomic factors obscures a pressing reality: that socioeconomic deprivation does not drive one to choose assisted dying, but it does increase one's mortality (Downar et al. 2023). In other words, all the time spent making rhetorical claims against assisted dying and how it fosters the killing of the socioeconomically deprived is time not spent actually attempting to fix what is really happening, which is that socioeconomic deprivation drives mortality up for those concerned.

We have established that, at the very least, class and rurality do indeed affect one's ability to access assisted dying. Recent research from Canada has also highlighted the link between vulnerable and marginalized populations, the elderly, the poor, the racialized, substance users and those with mental health issues, and barriers to access, calling for policymakers to ensure assisted dying is equally available to all citizens (Wright and Shaw 2018; Fortin et al. 2024).

Structural injustice should thus not be seen as a force that only cautions against assisted dying on grounds of devaluation or oppression; it is equally a force that, when left unaddressed, denies legitimate access to those entitled to it. A just society cannot be content merely with ensuring that no one is pressured into assisted dying; it must also ensure that structural barriers do not prevent rightful access. As I have

[12] This has led some to try to turn the argument on its head and argue that, in the context of assisted dying, the rich and white were actually the oppressed and structurally disadvantaged (Krag 2014). I do not engage with this perspective here, but such a line of argument seems at the very least suspicious in light of the definitions of structural injustice and vulnerability we have adopted here.

[13] Note that there is conflict evidence regarding the prevalence of women in assisted dying requestors and receivers. Empirical data in various contexts do not converge to a common result and in some cases, for instance in the Netherlands as far as psychiatric disorders are concerned, women seem to be overly represented compared to men. See Nicolini et al. (2021) for an in-depth discussion of this case.

argued throughout this final subsection, recognizing structural barriers to assisted dying should force a shift in both ethical reflection and policy design. It compels us to move beyond the familiar concern of preventing wrongful deaths toward a broader vision of justice that includes securing rightful access. If assisted dying is to be part of a just society's end-of-life care, then we must address the patterns of exclusion and inequality that continue to mark its practice. Future research and policy must take seriously the task of mapping and dismantling these barriers, ensuring that assisted dying, where legal, is not merely a formal right but a substantial one, available to all who are entitled to it. This is because we hold two-layered duty to do so: first, to ensure that all who are entitled to it have equal substantial access to assisted dying, and second to fight off displays of structural injustice whenever we detect them. Future research in this direction must however keep in mind that structural barriers to access and displays of structural injustice are best studied from the point of views of those at disadvantage themselves. At the time this book will come out, we know still little about where such barriers exist and how to best tackle them. We've seen the cases of rural communities and low-income groups, but one can speculate on the influence on access of any form of oppression that is not experienced by the culturally, socially, and politically privileged few that actually have recourse to assisted dying (Sikka 2021). Descriptive data, while indeed crucial, does not provide it with an extensive picture of structural barriers and how structural injustice interfaces with assisted dying, hence the vital need for more requestor-centered research on the matter in the future.

References

Anderson, Elizabeth S. 1999. What is the Point of Equality? *Ethics* 109:287–337.

Aragon, Corwin, and Alison M. Jaggar. 2018. Agency, Complicity, and the Responsibility to Resist Structural Injustice. *Journal of Social Philosophy* 49 (3): 439–460.

Atenasio, David. 2019. Blameless Participation in Structural Injustice. *Social Theory and Practice* 45 (2): 149–177.

Barutta, Joaquin, and Jochen Vollmann. 2015. Physician-Assisted Death with Limited Access to Palliative Care. *Journal of Medical Ethics* 41:652–654.

Battin, Margaret P., Agnes Van der Heide, Linda Ganzini, Gerrit Van der Wal, and Bregje D. Onwuteaka-Philipsen. 2007. Legal Physician-Assisted dying in Oregon and the Netherlands: Evidence Concerning the Impact on Patients in "Vulnerable" Groups. *Journal of Medical Ethics* 33 (10): 591–597.

Bérubé, Anouk, Diane Tapp, Sophie Dupéré, et al. 2025. Do Socioeconomic Factors Influence Knowledge, Attitudes, and Representations of End-of-Life Practices? A Cross-Sectional Study. *Journal of Palliative Care* 40 (2): 152–161.

Browne, Jude. 2024. The Political Implication of the 'Untraceability' of Structural Injustice. *Contemporary Political Theory* 23:43–65.

Chambaere, Kenneth, and Jan L. Bernheim. 2015. Does Legal Physician-Assisted Dying Impede Development of Palliative Care? The Belgian and Benelux Experience. *Journal of Medical Ethics* 41:657–660.

Colburn, Ben. 2025. Palliative Care-Based Arguments Against Assisted Dying. *Bioethics* 39:187–194.

Collins, Andrew, and Brendan Leier. 2017. Can Medical Assistance in Dying Harm Rural and Remote Palliative Care in Canada? *Canadian Family Physician* 63:189–190.

Della Croce, Yoann. 2020. Taking Oppression Seriously: Disability and Physician-Assisted Suicide. *Bioethica Forum* 13 (1): 22–27.

Downar, James, Susan MacDonald, and Sandy Buchman. 2023. Medical Assisted in Dying, Palliative Care, Safety, and Structural Vulnerability. *Journal of Palliative Medicine* 26 (9): 1175–1179.

Dworkin, Ronald. 1994. *Life's Dominion: An Argument About Abortion, Euthanasia, and Individual Freedom*. Vintage.

———. 2000. *Sovereign Virtue: The Theory and Practice of Equality*. Harvard University Press.

Dworkin, Ronald, Thomas Nagel, Robert Nozick, John Rawls, Thomas Scanlon, and Judith J. Thomson. 1997. Assisted Suicide: The Philosophers' Brief. *New York Review of Books* 27:41–47.

Emanuel, Ezekiel J., Bregje D. Onwuteaka-Philipsen, John W. Urwin, and Joachim Cohen. 2016. Attitudes and Practices of Euthanasia and Physician-Assisted Suicide in the United States, Canada, and Europe. *JAMA* 316 (1): 79–90.

Fortin, Sylvie, Sabrina Lessard and Marie-Eve Samson. 2024. "Diversity and Access to Palliative Care and Medical Assistance in Dying in an Urban Setting." *OMEGA – Journal of Death and Dying* 0 (0): 1-20.

French, Maddy, Lorraine Hansford, and Tess Moeke-Maxwell. 2023. Reflecting on Choices and Responsibility in Palliative Care in the Context of Social Disadvantage. *Palliative Care and Social Practice* 17:1–10.

Ganzini, Linda. 2014. Lessons from Legalized Physician-Assisted Death in Oregon and Washington. In *Palliative Care and Ethics*, ed. Timothy E. Quill and Franklin G. Miller. Oxford University Press.

Gerson, Mila, Gitten H. Koksik, Naomi Richards, Lars J. Matersvedt, and David Clark. 2020. The Relationship of Palliative Care with Assisted Dying Where Assisted Dying is Lawful: A Systematic Scoping Review of the Literature. *Journal of Pain and Symptom Management* 59 (6): 1287–1303.

Gyngell, Christopher, and Thomas Douglas. 2018. Selecting Against Disability: The Liberal Eugenic Challenge and the Argument from Cognitive Diversity. *Journal of Applied Philosophy* 33 (4): 319–340.

Hayward, Clarissa R. 2017. Responsibility and Ignorance: On Dismantling Structural Injustice. *The Journal of Politics* 79 (2): 396–408.

Hofmann, Bjorn. 2016. 'You are Inferior!' Revisiting the Expressivist Argument. *Bioethics* 31:505–514.

Kious, Brent. 2024. Hard Choices: How Does Injustice Affect the Ethics of Medical Aid in Dying? *Cambridge Quarterly of Healthcare Ethics* 33 (3): 413–424.

Krag, Erik. 2014. Rich, White, and Vulnerable: Rethinking Oppressive Socialization in the Euthanasia Debate. *The Journal of Medicine and Philosophy* 39 (4): 406–429.

Lippert-Rasmussen, Kasper. 2014. *Born Free and Equal? A Philosophical Inquiry into the Nature of Discrimination*. Oxford University Press.

Martin, Angela K., Nicolas Tavaglione, and Samia Hurst. 2014. Resolving the Conflict: Clarifying 'Vulnerability' in Health Care Ethics. *Kennedy Institute of Ethics Journal* 24 (1): 51–72.

Mathews, Jean J., David Hausner, Jonathan Avery, Breffni Hannon, Camilla Zimmermann, and Ahmed al-Awamer. 2021. The Impact of Medical Assistance in Dying on Palliative Care: A Qualitative Study. *Palliative Medicine* 35 (2): 447–454.

McKeown, Maeve. 2021. Structural Injustice. *Philosophy Compass* 16 (7): e12757.

Nicolini, Marie E., Chris Gastmans, and Scott Y. H. Kim. 2021. Psychiatric Euthanasia, Suicide and the Role of Gender. *The British Journal of Psychiatry* 220 (1): 10–13.

Reimer-Kirkham, Sheryl, Kelli Stajduhar, Bernie Pauly, et al. 2016. Death is a Social Justice Issue: Perspectives on Equity-Informed Palliative Care. *Advances in Nursing Science* 39 (4): 293–307.

Sangiovanni, Andrea. 2018. Structural Injustice and Individual Responsibility. *Journal of Social Philosophy* 49 (3): 461–483.

Schemmel, Christian. 2012. Distributive and Relational Equality. *Politics, Philosophy and Economics* 11 (2): 123–148.

Schiff, Jade. 2024. Structural Injustice and the Two Faces of Vulnerability. In *What is Structural Injustice?* ed. Maeve McKeown and Jude Browne. Oxford University Press.

Sikka, Tina. 2021. Barriers to Access: A Feminist Analysis of Medically Assisted Dying and the Experience of Marginalized Groups. *OMEGA – Journal of Death and Dying* 84 (1): 4–27.

Steck, Nicole, Matthias Egger, Maud Maessen, Thomas Reisch, and Marcel Zwahlen. 2013. Euthanasia and Assisted Suicide in Selected European Countries and US States. *Medical Care* 51 (10): 938–944.

Steck, Nicole, Christophe Junker, Maud Maessen, Thomas Reisch, Marcel Zwahlen, and Matthias Egger. 2014. Suicide Assisted by Right-to-Die Associations: A Population Based Cohort Study. *International Journal of Epidemiology* 43 (2): 614–622.

Steck, Nicole, Matthias Egger, and Marcel Zwahlen. 2016. Assisted and Unassisted Suicide in Men and Women: Longitudinal Study of the Swiss Population. *The British Journal of Psychiatry* 208 (5): 484–490.

Steck, Nicole, Christohpe Junker, and Marcel Zwahlen. 2018. Increase in Assisted Suicide in Switzerland: Did the Socioeconomic Predictors Change? Results from the Swiss National Cohort. *BMJ Open* 8:e020992.

Tavaglione, Nicolas, Angela K. Martin, Nathalie Mezger, et al. 2016. Fleshing out Vulnerability. *Bioethics* 29 (2): 98–107.

Voigt, Kristin. 2018. Relational Equality and the Expressive Dimension of State Action. *Social Theory and Practice* 44 (3): 437–467.

Wiebe, Kayla, and Amy Mullin. 2023. Choosing Death in Unjust Conditions: Hope, Autonomy, and Harm Reduction. *Journal of Medical Ethics* 50:407–412.

Wolff, Jonathan. 2024. Structural Harm, Structural Injustice, Structural Repair. In *What is Structural Injustice?* ed. Maeve McKeown and Jude Browne. Oxford University Press.

Wright, Alysia, and Jessica Shaw. 2018. The Spectrum of End-of-Life Care: An Argument for Access to Medical Assistance in Dying for Vulnerable Populations. *Medicine, Health Care and Philosophy* 22:211–219.

Young, Iris M. 1990. *Justice and the Politics of Difference*. Princeton University Press.

———. 2011. *Responsibility for Justice*. Oxford University Press.

Chapter 4
Conclusion: Politicizing Assisted Dying

Abstract This final chapter addresses the theoretical relationship between bioethics and political theory and argues that despite the common division of labor that separates the two disciplines, they are in fact much more interconnected than it might seem at first glance. I contend that accounting for the political conditions in which the subjects of bioethics are usually discussed is crucial to exhaustive normative analysis, as matters that pertain to the political, such as structural injustice, demand political answers. Following this line of thought, I then argue for a politicization of assisted dying, that is a proper acknowledgement of how a society's political background, in this case again structural injustice, inevitably interfaces with assisted dying policies and their effects. I conclude by laying out some potential future research avenues to that end.

Keywords Politicization · Bioethics · Political theory · Justice · Structural injustice · Assisted dying

We have now analyzed the extent to which specific instances of unequal access to assisted dying are discriminatory and how some of those discriminatory cases are less morally justifiable than others, chiefly when discrimination is not enshrined in law and intentional but rather pervasive and rooted in bias and systemic indifference. We have then reviewed the role of structural injustice in some of the arguments against assisted dying and showed that structural injustice may also be relevant in arguing for the expansion of assisted dying boundaries for it can create structural barriers that hinder access to the already structurally disadvantaged. As it tackles the subjects of discrimination and structural injustice, this book naturally holds a distinctively *political* undertone, insofar as those phenomena are distinctively political in nature. These political underpinnings have unfortunately long remained unaddressed, most likely due to an unfortunate division of labor between bioethics and political theory, as assisted dying is largely thought to be an issue of relevance for the former and not the latter. Hopefully we've successfully established that looking at the ethics of assisted dying from the viewpoint of political theory proves illuminating. What needs to be made absolutely clear is that those problems

Y. Della Croce, *Assisted Dying, Discrimination, and Structural Injustice*, SpringerBriefs in Ethics, https://doi.org/10.1007/978-3-032-11093-0_4

of political nature will require political solutions, hence why I want to end this short book with a call for further *politicization* of assisted dying.

I wish to start off this conclusion by arguing that the distinction between those two disciplines is not, and has never been, as clear-cut as it may seem at first glance, for two main reasons. First, bioethics has always been and is increasingly recognized to be heavily political in its endeavors, both in its subjects and in its way of addressing them. Second, both disciplines aim at informing, at least to a significant extent, public policy on the same grounds of justice, equality, liberty, legitimacy, and so on. As far as the ethics of public policy are concerned, both disciplines have significantly more in common than what is usually perceived. Through this short exposition, I wish to argue for the similarity of bioethics and political theory not based on their supposed resemblance as both belonging to the realm of applied ethics[1] but rather on their shared political components, which this book seeks to bring to the fore. Some scholars have gone so far to argue that the subject of medical ethics is essentially the subject of political philosophy, and that medical ethics ought to be considered as a subfield of political philosophy, largely because the ethical frameworks adopted by bioethicists to solve the moral controversies they are faced with are inevitably informed by the frameworks provided by normative political theory (Ezekiel 1991).

Bioethics is and has always been political, even though the so-called "politicization" of bioethics, better understood as the acknowledgement of the political component of bioethics, has only been put forward relatively recently in the literature (Bishop and Jotterand 2006; Brown 2009; Häyry 2022). The political nature of bioethics is to be found mostly in the ways in which it both shapes and is shaped by institutional, social, and political structures, as well as in the inherent divisive character of its topics and how the moral judgements citizens create about them reflect their own conceptions of the good; it is as such no different from the traditional problems of political theory that seek to arbitrate between particular conceptions of the good and the principles of justice agreed by all (Rawls 1993). The subject of bioethics is political as well as embedded in socio-political structures. These two dimensions have been referred to as respectively the *liberal* and *social* conceptions of bioethics, with the latter possessing an added normative component: in addition to being socially and politically situated, the task of bioethics is both to elucidate the socio-political structural features in which scientific and technological progress as well as policies concerning the biological functioning of citizens unfold, and highlight how such advances and policies impact or will impact citizens in terms of injustices (Weil-Dubuc 2018).[2] Let us first start with a short presentation of the key features of the liberal conception of bioethics.

---

[1] There moreover have been voices arising from both disciplines that have argued that neither should be simply considered as exercises in applied ethics. For political theory see part I of (Rawls 2001) and for bioethics see (Hedgecoe 2004; Wilson 2013).

[2] Weil-Dubuc further identifies a third kind of political conception of bioethics, namely *conservative*, according to which bioethics should take the form of a watchman whose goal is to set moral

In the words of Gutmann and Thompson, bioethics was built on conflictual and contentious moral issues, of which abortion or assisted suicide are only small illustrations (Gutmann and Thompson 1997). Different and competing conceptions of the good collide and conflict all the time in bioethical debates and yet one policy ought to be adopted despite these seemingly irreconcilable views: this ought to be done not according to some empirically verifiable truth but through normative preference, however this is to be defined (Gregg 2022). This problem is a staple and central point of concern for liberal political theory, as heavily stressed by Rawls in his *Political Liberalism*. Rawls refers to this idea as the fact of pluralism (or value pluralism), defined as the presence of a variety of deeply held comprehensive doctrines about the nature of the good (Rawls 1993). This fact is intrinsic to contemporary liberal democracies and one of the main tasks of liberal theory is to accommodate this pluralism while promoting principles of justice acceptable by all. Doing so ideally requires some form of neutrality towards the nature of the good on the part of the State as well as procedurally legitimate mechanisms to decide which policies ought to be adopted and which ought to be discarded, hence the need for a *political* conception of justice. These elements are ever-present in bioethics, where moral disagreement is fueled by overarching and ultimately irreconcilable conceptions of the good. This in turn has led some scholars to urge for the acknowledgement of the political nature of bioethical decision-making and the elaboration of fair deliberative or procedural mechanisms to ensure legitimacy and acceptability from citizens (Della Croce et al. 2021; Gregg 2022; Gutmann and Thompson 1997).

The recognition of the influence of particularistic worldviews in bioethical reasoning has also given birth to another epistemological strand in bioethics, namely *critical* bioethics, where researchers are urged to situate their reasoning and acknowledge how their moral intuitions are informed by their personal and cultural experiences and how they relate to broader socio-political power relations (Árnason 2015; Hedgecoe 2004; Ives and Dunn 2010). The influence of this second strand further displays itself in bioethics' growing interest and concern about the structural features, particularly injustices, of the socio political environments in which health policies, scientific progress and medical technologies unravel and which they ultimately shape. While this concern with structural injustices is not in itself new and has been especially present at the interface of bioethics and disability studies for quite some time (Shakespeare 2019), the Covid-19 pandemic constituted a sharp turn in the way bioethicists engage with the subject (Churchill et al. 2020). The pandemic along with the growing reach of the Black Lives Matter movement have for instance drawn significant and welcome attention to the issues surrounding race in both bioethics as a discipline and health policies (Russell 2022), for instance in how they pertain to health disparities (Hutler 2022), vaccine access (Gayle and Childress 2021), the insufficient lack of interest in the roles race plays and has played in bioethics (Fletcher et al. 2022) or the subsequent lack of research funding

---

boundaries not to be crossed by technological progress. This is of no relevance for the present endeavor.

caused by this lack of interest (Fabi and Goldberg 2022). Others have seized the opportunity to emphasize the importance of social class in bioethics and urged for a lesser bourgeois bioethics blind to socio-economic inequalities (McCarthy 2021) or the active engagement of bioethicists in addressing structural injustices through the political commitment of addressing and amplifying the variety and multiplicity of voices affected by injustice and oppression (Macneill et al. 2021). This rising concern has furthermore led some to argue for the active highlighting and deconstruction of structural injustices in both clinical and educational settings, stressing the importance to consider structural injustices for both the medical practitioner and student as well as the bioethicist. This concern with the political emphasizes the aforementioned social conception of bioethics and the growing need for an understanding of bioethics as a socially and politically embedded discipline, both in its subjects and practice. This second aspect of politically charged bioethics transpires throughout this book. Indeed, this short book chiefly aimed at considering the socio-political structures which engender oppression and injustice and analyze how they display themselves in the cases of unjust inequalities of access to assisted dying, particularly through structural barriers.

As we have seen throughout Chap. 2, inequalities of access to assisted dying are manifold and not necessarily unjustified, even if they conceptually qualify as discriminatory. There may be a variety of reasons why some individuals are granted access to medical assistance in dying while denying it to others. Even the liberally inclined theorist in favor of assisted dying will usually not, for instance, welcome a legislation that completely fails to guarantee a minimum degree of autonomy in the requestor's choice to die. Except for libertarians, we tend to think that a minimal safety net is still required in those cases, and that it is the State's duty to implement policies that protect the most vulnerable against pressured choices or any form of choice that is not truly theirs (Mason 1990; Scoccia 1990). The extent to which this protection ought to reach is still heavily debated in the literature, and there is no substantial consensus on whether citizens suffering from depression, other mental illnesses, dementia, or existential suffering should be granted access just like the rest of the population or not. There are indeed cases where safeguards, if poorly thought out, can end up acting as structural barriers which unjustly prevent access for those already structurally disadvantaged (Downar et al. 2024). The best way to properly think about safeguards as well as guaranteeing justice in access to assisted dying is to take structural injustices into account and carefully analyze how they may lead to discriminatory practices. In addition to the illuminating nature of the concept of structural injustice in providing a framework for the cases we analyzed here, the beneficial relation also goes the other way, for in order to gain a greater understanding and finer conceptualization of structural injustice, one also needs to look at specific phenomena in which structural injustice displays itself (Sankaran 2021). Furthermore, proceeding in this fashion also allows for the unmasking of social processes of exclusion and marginalization, which in turn helps reinforce the political significance of both the specific instances of injustice and their structural counterpart of which they are a display (Haslanger 2012).

While death is certainly an event of capital importance in any individual's life and choosing one's way to die is one of the most intimate choices one can make, it does not follow that death is a strictly private matter. Death is, and has always been, deeply political in the sense that the sociopolitical and economic conditions one lives their life in will inevitably affect one's end-of-life, be it through how they impact their mortality (Antonovsky 1967; Elo 2009), access to palliative care (Kessler et al. 2005; Lewis et al. 2011; Bradshaw et al. 2024), or as it has been emphasized throughout this book, access to assisted dying. A politicization of assisted dying demands that we think about assisted dying less in terms of a narrow sense of applied ethics and more in terms of larger sociopolitical phenomena in which it takes place. It requires of us to think of assisted dying as an issue where pre-existing relations of power display themselves in a variety of ways, of which inequality of access is but one. It also requires of us to inform our normative prescriptions regarding assisted dying with politically infused ideas instead of attempting to solve its issues solely from the point of view of a narrow vision of applied ethics that foregoes the structural. What we have seen throughout this short book falls within the realm of conceptual clarification regarding the role of structural injustice and discrimination in assisted dying. While more work in this direction is definitely needed in order to better understand the many, so far blurry realities we have highlighted throughout this investigation, even more pressing work is needed in finding adequate political solutions to currents state of injustice. We have touched very little on what can be done to combat structural injustice in the context of assisted dying, and for one good reason: doing so would require another separate volume given the depth of the subject. An often-given answer to the problem of structural injustice is the substantial inclusion of the disadvantaged in decision-making that directly affects them. Thus, perhaps the way to go, or at least a way to go, is to start thinking of the ethics of assisted dying in procedural terms as the best way to alleviate the weight of structural injustice. All of these are examples of questions that arise from properly politicization assisted dying and overcoming the division of labor between bioethics and political theory by recognizing that both are in fact concerned with the same questions. Unfortunately, these kinds of endeavors cannot be undertaken here, yet one can hope that it will be in the near future. This short book has attempted to provide a small steppingstone towards this end.

References

Antonovsky, Aaron. 1967. Social Class, Life Expectancy and Overall Mortality. *The Milbank Memorial Fund Quarterly* 45 (2): 31–73.

Árnason, Vilhjalmur. 2015. Toward Critical Bioethics. *Cambridge Quarterly of Healthcare Ethics* 24 (2): 154–164.

Bishop, Jeffrey P., and Fabrice Jotterand. 2006. Bioethics as Biopolitics. *Journal of Medicine and Philosophy* 31 (3): 205–212.

Bradshaw, Andy, Naomi Richards, Jamilla A. Hussain, and Joanna M. Davies. 2024. We Need to Talk about Social Class: Why Theories of Social Class Matter for Understanding Inequities in Palliative and End-of-Life Care. *Palliative Medicine* 38 (10): 1076–1078.

Brown, Mark B. 2009. Three Ways to Politicize Bioethics. *The American Journal of Bioethics* 9 (2): 43–54.

Churchill, Larry R., Nancy M. P. King, and Gail E. Henderson. 2020. The Future of Bioethics: It Shouldn't Take a Pandemic. *Hastings Center Report* 50 (3): 54–56.

Della Croce, Yoann, Matteo Gianni, and Valeria Marino. 2021. What's True in Truth and Reconciliation? Why Epistemic Justice is of Paramount Importance in Addressing Structural Racism in Healthcare. *The American Journal of Bioethics* 21 (3): 92–94.

Downar, James, Eliana Close, Jessica E. Young, and Ben P. White. 2024. Assisted Dying: Balancing Safety with Access. *British Medical Journal* 387:q2382.

Elo, Irma T. 2009. Social Class Differentials in Health and Mortality: Patterns and Explanations in Comparative Perspective. *Annual Review of Sociology* 35:553–572.

Ezekiel, Emanuel. 1991. *The Ends of Human Life: Medical Ethics in a Liberal Polity*. Harvard University Press.

Fabi, Rachel, and Daniel S. Goldberg. 2022. Bioethics, (Funding) Priorities, and the Perpetuation of Injustice. *The American Journal of Bioethics* 22 (1): 6–13.

Fletcher, Faith E., Keisha S. Ray, Virginia S. Brown, and Patrick T. Smith. 2022. Addressing Anti-Black Racism in Bioethics: Responding to the Call. *Hastings Center Report* 52 (S1): S3–S11.

Gayle, Helene D., and James F. Childress. 2021. Race, Racism, and Structural Injustice: Equitable Allocation and Distribution of Vaccines for the COVID-19. *The American Journal of Bioethics* 21 (3): 4–7.

Gregg, Benjamin. 2022. Political Bioethics. *The Journal of Medicine and Philosophy: A Forum for Bioethics and Philosophy of Medicine* 47 (4): 516–529.

Gutmann, Amy, and Dennis Thompson. 1997. Deliberating About Bioethics. *Hastings Center Report* 27 (3): 38–41.

Haslanger, Sally. 2012. *Resisting Reality: Social Construction and Social Critique*. Oxford University Press.

Häyry, Matti. 2022. *Roles of Justice in Bioethics*. Cambridge University Press.

Hedgecoe, Adam M. 2004. Critical Bioethics: Beyond the Social Science Critique of Applied Ethics. *Bioethics* 18 (2): 120–143.

Hutler, Brian. 2022. Causation and Injustice: Locating the Injustice of Racial and Ethnic Health Disparities. *Bioethics* 36 (3): 260–266.

Ives, Jonathan, and Michael Dunn. 2010. Who's Arguing? A Call for Reflexivity in Bioethics. *Bioethics* 24 (5): 256–265.

Kessler, David, Tim J. Peters, Liz Lee, and Susie Parr. 2005. Social Class and Access to Specialist Palliative Care Services. *Palliative Medicine* 19:105–110.

Lewis, M. Joanne, Michelle DiGiacomo, David C. Currow, and Patricia M. Davidson. 2011. Dying in the Margins: Understanding Palliative Care and Socioeconomic Deprivation in the Developed World. *Journal of Pain and Symptom Management* 42 (1): 105–118.

Macneill, Paul, Christopher F. C. Jordens, Deborah Zion, and Angus Dawson. 2021. Bioethics as Engaged Activity. *The American Journal of Bioethics* 21 (2): 64–66.

Mason, Andrew D. 1990. Autonomy, Liberalism and State Neutrality. *The Philosophical Quarterly* 40 (161): 433–452.

McCarthy, Michael. 2021. Beyond a Bourgeois Bioethics. *Journal of the Society of Christian Ethics* 41 (1): 73–88.

Rawls, John. 1993. *Political Liberalism*. Columbia University Press.

———. 2001. *Justice as Fairness: A Restatement*. Harvard University Press.

Russell, Camisha. 2022. Meeting the Moment: Bioethics in the Time of Black Lives Matter. *The American Journal of Bioethics* 22 (3): 9–21.

Sankaran, Kirun. 2021. "Structural Injustice" as an Analytical Tool. *Philosophy Compass* 16 (10): e12780.

Scoccia, D. 1990. Paternalism and Respect for Autonomy. *Ethics 100* (2): 318–334.
Shakespeare, Tom. 2019. When the Political Becomes Personal: Reflecting on Disability Bioethics. *Bioethics* 33 (8): 914–921.
Weil-Dubuc, Paul-Loup. 2018. Moral, sociétal ou social: Quel rôle politique pour la bioéthique? In *Traité de bioéthique*, ed. Emmanuel Hirsch. Erès Espace éthique–Poche.
Wilson, Duncan. 2013. What Can History do for Bioethics? *Bioethics* 27 (4): 215–223.

GPSR Compliance
The European Union's (EU) General Product Safety Regulation (GPSR) is a set of rules that requires consumer products to be safe and our obligations to ensure this.

If you have any concerns about our products, you can contact us on

ProductSafety@springernature.com

In case Publisher is established outside the EU, the EU authorized representative is:

Springer Nature Customer Service Center GmbH
Europaplatz 3
69115 Heidelberg, Germany

www.ingramcontent.com/pod-product-compliance
Lightning Source LLC
Chambersburg PA
CBHW070852091225
36512CB00013B/50
* 9 7 8 3 0 3 2 1 1 0 9 2 3 *